The Effectiveness of European Union Environmental Policy

Wyn Grant
Professor of Politics
University of Warwick

Duncan Matthews
Lecturer in the School of Law
University of Warwick

and

Peter Newell
Research Fellow
Institute of Development Studies
University of Sussex

First published in Great Britain 2000 by
MACMILLAN PRESS LTD
Houndmills, Basingstoke, Hampshire RG21 6XS and London
Companies and representatives throughout the world

A catalogue record for this book is available from the British Library.

ISBN 0–333–73065–8 hardcover
ISBN 0–333–73066–6 paperback

First published in the United States of America 2000 by
ST. MARTIN'S PRESS, LLC,
Scholarly and Reference Division,
175 Fifth Avenue, New York, N.Y. 10010

ISBN: 0–333–73065–8 (Cloth)
0–333–73066–6 (Paper)

Library of Congress Cataloging-in-Publication Data
Grant, Wyn.
The effectiveness of European union environmental policy / Wyn Grant, Duncan Matthews, and Peter Newell.
p. cm.
Includes bibliographical references and index.
ISBN 0–333–73065–8 — ISBN 0–333–73066–6 (pbk.)
1. Environmental policy—Economic aspects—European Union countries. 2. Environmental protection—European Union countries. 3. Environmental law––European Union countries. I. Matthews, Duncan, 1965– II. Newell, Peter (Peter John) III. Title.

HC240.9.E5 G728 2000
333.7'094—dc21

00–062611

This book is printed on paper suitable for recycling and made from fully managed and sustained forest sources.

10 9 8 7 6 5 4 3 2 1
09 08 07 06 05 04 03 02 01 00

Printed and bound in Great Britain by
Antony Rowe Ltd, Chippenham, Wiltshire

Contents

Preface

This book results from collaboration which started when all three authors were in the Department of Politics and International Studies at the University of Warwick, although Peter Newell is now a Research Fellow at the Institute of Development Studies at the University of Sussex. The early phases of the book involved collaboration with Steven Kennedy at Macmillan and we are grateful to him for his editorial advice and assistance.

We have followed the convention established by William and Helen Wallace in their edited book *Policy-Making in the European Union* and for the sake of simplicity referred to the European Union, except where the references are specific to pre-Maastricht events.

Duncan Matthews would like to express his gratitude to the Economic and Social Research Council for providing generous financial support within the ESRC Single European Market Programme (award no. R000233673) and the ESRC Global Environmental Change Programme (award no. L320223011). The results of research funded by the ESRC form the basis of Chapter 3 and Chapter 6. The latter chapter also includes a discussion of implementation and enforcement problems originally raised in the context of a Department for Education and Employment (DfEE) research project on 'The Implications of the Evolution of European Integration for UK Labour Markets', the results of which were published in the DfEE research series report no. 74 (March 1996). Duncan Matthews is also grateful to John Pickering for his constant moral support and expert advice on the economics of environmental policy and to former colleagues at the National Institute of Economic and Social Research for providing a genuinely supportive research community. Finally, he would like to thank the ESRC Centre for the Study of Globalisation and Regionalisation at the University of Warwick for the opportunity to compile this work in 1998.

Peter Newell would like to thank the following for providing useful advice about literatures and sources of information:

Sonja-Boehmer-Christiansen, Ute Collier, Dave Humphreys, Ian Rowlands and John Vogler. Charlie Dannreuther and Ben Rosamond deserve a special mention for being helpful and enthusiastic well beyond the call of duty, and for being great friends. Thanks are also due to former colleagues at Climate Network Europe for many useful conversations and access to all their archives, to Liam Salter from CNE for helpful comments on the climate change chapter and to Linda Bateman for her administrative support.

Wyn Grant was able to work on this book while a Marshall–Monnet Fellow at the EU Center of the University of Washington in the spring quarter 1999 and in particular would like to thank the Center's director, Professor John Keeler. Professor William Paterson provided helpful comments on an early version of the plan for the book.

Any errors in the book are, of course, the responsibility of the authors and not of any of the individuals mentioned in this preface. The authors would also like to thank their partners for their forbearance and support.

WYN GRANT
DUNCAN MATTHEWS
PETER NEWELL

Introduction

In an era when awareness of the risks of government failure has increased, it is more important than ever that public policies should be effective. What this requires is that a policy should achieve its objectives with the minimum of undesirable side-effect; with as little public expenditure as possible; and with as few additional costs for those impacted by the policy as is feasible. Achieving policy effectiveness is particularly important in the case of environmental policy where the policies are less well entrenched than older public policies and the benefits tend to be apparent only in the longer run.

There has been an extensive debate about the definition and measurement of effectiveness within the literature on international relations (for a recent example, see Jordan, 1998). However, in so far as this debate has been concerned with 'The most simple criterion of success . . . whether the goals embodied in the policy in question are achieved' (Jordan, 1998, p. 24), it does not really provide a framework for exploring the complexities of environmental policy in the European Union. We have therefore evolved a framework for testing policy effectiveness based on a public policy approach.

Special difficulties arise in the case of the European Union policy process. Wallace (1996, p. 29) argues 'that tests of policy effectiveness play a particularly important part in the European policy process'. This is because European policies have to establish their credibility as more efficient alternatives to national-level policies and because the legitimacy of European policies is fragile. Effective policies can help to build up new bases of support, but this is dependent on results.

Wallace (1996, p. 30) sets out eight criteria for measuring policy effectiveness. These may be regarded as optimal criteria which are unlikely to be met by any policy, let alone environmental policy. Indeed, as will become evident, it is the least stringent criterion towards the end of the list that come the closest to being met. Nevertheless, they provide a useful starting point for analysis.

The first criterion is that policy authority should be clearly established at the European level. For a long time, this was not the case with environmental policy – it was not mentioned in the Treaty of Rome and progress had to be justified in terms of the creation of the single market or the achievement of new goals under what was then Article 235. The Single European Act and the Treaty on European Union remedied this particular defect. Most environmental policy is now clearly promulgated at the European level, even though it is usually implemented at national, subnational and local levels.

The second requirement is that the policy is 'based on rules that can be and are backed by tough European law' (Wallace, 1996, p. 30). This criterion is met in the sense that at least major companies regard European environmental directives as something they have to take into account and comply with, although enforcement standards may vary between member states. The policy area is also one which, through the actions of the European Court of Justice, has been active in developing the scope and applicability of the law as is discussed in Chapter 1. Moreover, EU directives may give local authorities an authoritative basis for effective action they would otherwise lack (Grant, Perl and Knoepfel, 1999).

The third criterion is that the policy should be backed by resources which are distributed at the European level, thus providing incentive structures. In Lowi's (1964) terms, environmental policy is essentially regulatory rather than distributive or redistributive. It does not create some clientele of beneficiaries who agitate for the continuation and the development of the policy. This does not matter, however, if we accept that a characteristic of the 'emerging European polity [is] the significance of regulation as the main instrument of public power in the Union' (Laffan and Shackleton, 1996, p. 74).

The fourth test is one of the hardest to meet, but also one of the most important – that the 'policy may change behaviour by relevant actors' (Wallace, 1996, p. 30). The European Community has produced an impressive pile of paperwork in terms of directives and environmental action programmes since it started to develop what might be called an environmental policy in the early 1970s. What is important in terms of policy effectiveness, however, is not the output of the legislative factory, but whether those directives have been converted into policy measures which are enforced in a way that leads to effective environmental outcomes.

A central concern of this book is thus with the implementation and enforcement of environmental policy. It is the subject of a separate chapter, but is also a theme that runs through the analysis of individual policy areas. Effective implementation and enforcement may not by itself change the behaviour of relevant actors such as businesses, but it is an essential precondition. Implementation is in the hands of the member states and those in charge of the activity being regulated. It therefore depends on the active cooperation of the member state, its administrative capacity and the cooperation of those being regulated before behaviour can start to be changed.

Wallace's fifth criterion is that policy needs to be based on innovative ideas developed at the European level, which means that effectiveness is based on remedies that enjoy intellectual integrity. Although this book does not consider in detail how the ideas which inform environmental policy emerge, our discussion does show how non-governmental organisations contribute to the policy debate and how the development of policy in particular areas such as ground-level air pollution depends on technical knowledge about environmental phenomena and their impact on human health. Regulatory policy requires a thorough and expert knowledge of the regulated activities. This creates a central role for experts in the policy process.

A more general question that underpins the book is whether the development of the issue area has led the EU to think afresh about the ways in which it goes about its business, or whether it has merely been captured within existing frameworks of understanding and doing things. Has anything changed about the way in which the EU operates as a result of the emergence of environmental policy, or has environmental policy been obliged to operate within rules of the game that are not appropriate to the issue area? In this way, we seek to raise broader questions about the integration process.

The sixth requirement is that policy should be based on an equilibrium point 'where everyone relevant is as well off as possible' (Wallace, 1996, p. 30). Whether this requirement is actually met in European environmental policy is difficult to assess in practice. The environmental benefits of policy are generally spread across society, although the costs of environmental improvements may be borne disproportionately by one group – often business. However, much of the discussion in this book suggests that European environmental policy falls well short of what might be regarded as an optimal outcome. Frequently, it is the case that the policy that emerges is often the best bargain that can be negotiated politically at a given time.

That the seventh and eighth criteria are the easiest to meet tells us something about the condition of European environmental policy. The 'policy may not be excellent, but alternatives are worse' (Wallace, 1996, p. 30) is a revealing statement when one realises that it sits well with much of what constitutes European environmental policy. It may not effectively tackle global warming, but it represents more than is being tried elsewhere in the world; it may not bring us clean water, but at least it gives us cleaner water than we would otherwise have; and, while it may not stop ground-level air pollution breaching agreed standards, it may prevent any further deterioration.

Finally, 'policy effectiveness within the issue area may be questionable, but policy serves symbolic . . . goals' (Wallace, 1996, p. 30). Environmental policy is of symbolic importance in the EU for at least three reasons. First, it deals with a transboundary phenomenon which should be able to be dealt with most effectively at a European level. If the EU cannot develop an effective strategy for environmental management, it is unlikely to succeed in other policy arenas. Second, it is a policy which allows the EU to engage directly with the broad mass of its citizens.

Unlike agricultural policy, which directly affects only a small proportion of EU citizens as producers, environmental policy potentially affects the quality of everyone's life. Third, it is a set of issues on which the EU can take a distinctive stance in international negotiations, not least differentiating itself from the United States but in a more positive way than occurs in trade negotiations, where the EU's emphasis is often on seeking to defend market-distorting policies like the CAP. One would hope, however, that EU environmental policy serves substantive as well as symbolic goals. We shall explore the substance of EU environmental policy throughout this book and return to this issue of whether its impact is predominantly symbolic in the conclusion.

It is open to question how far there is something that can be called 'environmental policy'. This is perhaps why a definition of the term is often evaded. In one sense, a definition might seem unnecessary as the environment is that which surrounds us, the physical set of conditions in which we and other sentinent beings exist.

One might go on to argue that environmental policy is what the EU seeks to do to protect the physical environment. The EU, of course, produces regular environmental action programmes to guide its activities and there is a recognisable environmental policy community whose elements are reviewed in Chapters 1 and 2. On the whole, however, policy is not guided by some underlying conception of an ecological system which has to be viewed as a whole. Rather, policy is highly segmented with each aspect of policy proceeding at its own pace and in accordance with its own priorities. In part, this reflects the extent to which the policy-making process is dependent in terms of both specialist knowledge and legitimacy on the mobilisation of scientific expertise.

This characteristic segmentation of environmental policy is why a considerable portion of this book is devoted to the study of particular policy areas. In practice, water and air pollution have been major areas of activity, and each of these topics is the subject of a chapter. In the 1990s, the issue of global warming has risen to the top of the environmental policy agenda and this issue has an added interest because it involves the EU in the negotiation of an international policy regime.

Collier asserts (1997a, p. 1) that 'Environmental policy is considered to be one of the European Union's most successful policies', but what does this success consist of? Admittedly, from a position where there was no environmental policy at all, the EU has developed an extensive policy which has a base in the treaties, its own directorate-general and is embodied in several hundred directives and regulations. The key questions are 'How well overall is the environmental system of governance performing? In particular, how far is it producing high quality decisions that are well adapted to the problems at which they are directed?' (Weale, 1996, p. 609). If decision-making is often suboptimal, as Weale suggests, this reduces the chances of securing the desired outcomes. It is to the question of how the institutions of the EU take decisions about environmental policy that we turn to in Chapter 1.

1 Making Environmental Policy

What is environmental policy?

Most standard accounts of the European Union (EU) do not consider it necessary to define what is meant by environmental policy and equally most books on the subject launch into their discussion without considering what it is that is being assessed. In one sense, a definition may seem unnecessary: the environment is that which surrounds us, the physical set of conditions in which we and other sentient beings exist. One might go on to argue that environmental policy is what the EU seeks to do to protect the physical environment.

It is at least necessary to pause to consider what environmental policy consists of and what its boundaries are. Despite what is in many ways its global character, the issues on the environmental agenda differ from one part of the world to another. For example, in the Pacific North West in 1999 the three main environmental issues on either side of the border between the United States and Canada were: (i) salmon; (ii) whether native peoples should be allowed to exercise their treaty rights to hunt whales; and (iii) old growth forests. None of these issues has any particular resonance in the EU.

The policy process is segmented into a number of vertical compartments, but environmental policy is itself in turn highly segmented. Although the EU's environmental action plans represent an attempt to define a set of priorities and policy for the environment as a whole, in effect one has a set of distinct policies related to very specific objectives to the extent that it is difficult to talk of an overall environmental policy. In part, this reflects the extent to which the policy-making process is dependent on the mobilisation of scientific expertise. Someone who knows about the dispersal characteristics of an ozone plume from a metropolis may know very little about alternative models of global warming, and will almost certainly know very little about water pollution or toxic contaminants in the soil.

In practice within the European Union, 'Most activity has been in the area of water and air pollution' (Collier, 1997a, p. 2). This is understandable because it is where some of the most evident environmental problems have been whether the pollution of rivers such as the Rhine, the contamination of bathing beaches or increasing concern about the health risks arising from ground-level air pollution. In the 1990s, in the wake of the Rio Earth Summit of 1992, the issue of global warming has risen to the top of the environmental policy agenda and this issue has an added interest because it involves the European Union in the negotiation of an international policy regime (see Chapter 4). Another area of activity has been the issue of waste both in terms of waste minimisation through action on packaging, the encouragement of recycling and the treatment of sites that include toxic wastes.

By contrast, one area in which there has been less consistent activity is that of nature conservation and biodiversity, although the Wild Birds Protection Directive of 1979 was one of the early landmark directives. One of the problems in this area is that any attempt at effective action runs up against the Common Agricultural Policy which has encouraged chemically intensive forms of agriculture which are inimical to wildlife. A set of studies commissioned by the four countryside agencies in Britain argued that 'The current beef, dairy and sheepmeat regimes have encouraged farming systems to develop which have ignored the land's "natural carrying capacity"' (*Agra Europe*, 21 August 1998, p. EP/4). It is at the interface between agriculture and the environment that one becomes aware of some of the boundaries of environmental policy. For instance, animal protection and welfare policies are not generally seen as environmental policies in so far as they apply to the housing, care or transport of farm animals. By contrast, animals which generally live within the wild, such as birds, seals and whales, are seen as falling within the scope of environmental policy.

The development of environmental policy

There was no reference to environmental policy in the Treaty of Rome of 1957. It was not an issue in the period of postwar recovery, but as public interest in problems of pollution grew, it became increasingly apparent to European leaders that the European Community should involve itself in a set of problems that do not respect national boundaries. What is sometimes regarded as the EC's first environmental directive was passed in 1967 dealing with standards for classifying, packaging and labelling dangerous substances, but its real focus was on the facilitation of trade. Subsequent legislation built on this framework directive, notably the 6th Amendment of 1979 which provided for the pre-market control of hazardous chemicals. This might more genuinely be regarded as an environmental directive.

Nevertheless, Scott (1998, p. 4) has identified 150 separate pieces of environmental legislation adopted during the period 1967 to 1987 – the period up until the Single European Act finally introduced a new 'Environment' title into the Treaty. During this initial period of ad hoc, piecemeal expansion in EU environmental policy competence, the Commission proved creative in the use of Article 100, which allowed for the approximation of member state laws which directly affect the establishment or functioning of the common market, and Article 235, which allows for the adoption of Community measures where necessary to attain, in the course of the operation of the common market, one of the objectives of the Community where the Treaty has not provided the necessary powers.

This development of environmental competence was given a major impetus at the 1972 Paris Summit when the heads of government called upon the Commission to draw up an environmental policy and set up a directorate responsible for environmental protection. A step had already been taken in this direction with the formation of an Environment and Consumer Protection Unit in 1971. In 1972 this became an Environment and Consumer Protection Service with 15 staff members attached to the Industrial

Policy Directorate, DG III (McCormick, 1998, p. 192). In 1973 the Community published its first Environmental Action Programme which emphasised the harmonisation of national policies between member states. 'In 1981 a reorganization of the Commission in the light of Greek accession resulted in the transfer of environmental responsibilities from DG III to a reformulated DG XI' (McCormick, 1998, p. 193). There was not, however, a separate environment portfolio for a commissioner until the appointment of Carlo Ripa de Meana in 1989. (Although Stanley Clinton-Davis held a joint transport and the environment portfolio from the mid-1980s.)

However, it is important to be aware of the limits of what could be achieved at a time of severe economic turbulence characterised by the phenomenon of 'stagflation' (an economic condition characterised by minimal growth and high inflation). The environment Council met just twice a year in 1975 and 1980, compared with 14 and 15 meetings respectively for the agriculture Council (Hayes-Renshaw and Wallace, 1997, p. 30). The extensive diaries of Roy Jenkins on his period as president from 1977–81 make just one reference to environmental matters, in the context of the brusqueness of a British minister, but several references to problems about butter (Jenkins, 1989, p. 186):

> After the 'conversion' of the German government to a pro-environment stance in 1982 on the question of acidification, the pace of developments speeded up considerably, notably with the 1984 framework directive on the control of air pollution from large stationary sources and the passing of the environmental impact directive of 1985. (Weale, 1996, p. 597)

However, the development of environmental policy was handicapped by the lack of any basis in the treaties. Environmental measures had to rely on the harmonisation provisions of Article 100 or the general provisions of Article 235.

This deficiency was remedied by the Single European Act (SEA) of 1987 which provided, in Articles 130(r–t), a new treaty basis for decisions about the environment. This gave explicit recognition to the improvement of environmental quality as a legitimate Community objective in its own right. 'This meant that EC environmental policies need no longer be justified in terms of their contribution to economic integration' (Vogel, 1996, p. 125). The 'polluter pays' principle was recognised and the SEA stated that in harmonising national regulations the Community would take as a base a high level of environmental protection. A number of alternative decision-making routes were introduced involving a complex mixture of unanimity and qualified majority voting. However, the SEA gave a considerable impetus to environmental legislation. 'Between 1989 and 1991, the EC enacted more environmental legislation than in the previous twenty years combined' (Vogel, 1996, p. 127).

Environmental policy has moved on from simply tackling evident pollution problems such as those of the the Rhine or the North Sea, or ensuring that proper regulations are in place to control hazardous substances such as mercury, towards a greater emphasis on the 'precautionary principle' and to reviewing the coherence of policies which have developed on an ad hoc basis (see the chapter on water policy). The 'precautionary principle' has never been clearly defined, but environment commissioner Margot Wallström sees it as directing the EU to 'take action when the science is not clear, but where there is reasonable cause for concern' (*Agra Europe*, 4 February 2000, p. EP/6).

The historical emphasis in environmental policy was on 'command and control' regulation. However, an increasing realisation developed that this was a regulatory approach was inflexible and inefficient in the sense that it carried with it high transaction costs and a poor record in terms of securing the desired environmental outcomes. Uniform reduction targets ignore the reality that marginal costs of pollution reduction can vary from firm to firm or industry to industry. 'Command and control is not only an expensive approach to pollution reduction, but one which, according to many analysts, has also reached the limits of its environmental effectiveness' (Golub, 1998b, p. 4).

It was also seen as incompatible with the principle of subsidiarity. This led to a growing interest in New Environmental Policy Instruments (NEPIs) which was reflected in the Fifth Environmental Action Programme of 1992. NEPIs can be market based such as 'green taxes' or tradeable emissions permits or they can involve voluntary agreements and partnerships with business, along with ecolabels and ecoaudits. A review of progress in the area suggests that there is a need 'to reconsider how we judge the "effectiveness" of new instruments, which in turn raises important questions about proper environmental policy design' (Golub, 1998b, p. 23).

The fundamental problem is that three different and often incompatible aspects of effectiveness are being pursued simultaneously: cost savings, political legitimacy and environmental improvement. Critics of the new approach argue that the first two are pursued at the cost of the third. Command and control approaches are still relevant in some circumstances and are unlikely to be totally abandoned, so the question becomes what combination of old and new policy tools should be used (Golub, 1998b, pp. 23–4).

The use of a wider and more sophisticated range of policy instruments is one sign that the environmental policy arena in the EU has developed greater maturity, becoming proactive rather than reactive. However, there are also signs of a loss of impetus. 'Despite some consolidation and advance since 1992, the scale and pace of development have slowed down considerably, and some high profile measures have been stalled' (Weale, 1996, p. 609). A DG XI official commented to one of the authors, 'We are taking a step-by-step approach now, the carbon energy tax having failed, trying to be more pragmatic and less ambitious.'

What distinguishes environmental policy from other EU policy areas?

Anyone wishing to understand a particular policy sector in the EU would be well advised to start by looking at three actors: the relevant directorate-general; the Council of Ministers for the policy arena; and the interest groups associated with the policy. Next, the analyst might turn to the Court of Justice and the European Parliament to see if they had made any significant interventions in the policy process, probably paying more attention to the Court than the Parliament. They might also see if any policy disputes had had to be resolved by the European Council of heads of government.

This crude policy model would work well enough for the environmental arena and later in this and subsequent chapters we will consider the roles of these various actors in the formulation of European environmental policy, as well as considering the recently formed European Environment Agency. In terms of the institutions and processes involved, EU environmental policy is superficially not very different from any other policy area. However, one of the arguments of this book is that EU environmental policy does have a distinctive character in terms of such dimensions as the legitimacy of EU policy competence, the policy instruments available to achieve policy goals and implementation or enforcement problems associated with the achievement of those goals.

One potentially significant difference is that it is a younger and, in some respects, less well-entrenched policy sector than most others. In order to achieve an accepted status, it had to define itself in line with single market policy objectives. The SEA required in Article 132r(2) that 'environmental protection requirements shall be a component of the Community's other policies'. However, this wording was clearly insufficiently strong to overcome inbuilt institutional resistance from directorates-generals with different missions. The Treaty of Amsterdam contains a provision that 'Environmental protection requirements must be integrated into the definition and implementation of Community policies . . . in particular with a view to promoting sustainable development.'

The fact that such a provision has to be made does reveal some of the difficulties that environmental policy-makers encounter, particularly when they move into areas which impinge on the activities of other directors-general. This point is discussed further in the examination of the Environment Directorate-General and its relations with other DGs later in the chapter. For the moment, it is sufficient to note that environmental policy does not enjoy the same degree of institutional displacement in terms of priority in the overall policy agenda and the provision of administrative resources as areas such as agricultural policy or social policy. These are well-established policy areas with highly integrated policy networks which are at the forefront of the EU's activities. No president of the Commission could ignore them, whereas environmental policy was not a priority even for such an activist president as Jacques Delors.

There is also a sense in which the debate can be a relatively narrow one, conducted within an environmental policy paradigm which may be caricatured as a four-stage process: (i) identify problem; (ii) consult scientists to characterise problem and identify possible solutions; (iii) draft and negotiate directive; (iv) begin struggle to implement directive (meanwhile many years have passed). Jordan, Brouwer and Noble (1999, p. 384) found that the average period of time needed to adopt environmental legislation between 1967 and 1997 was 28 months. However, they took their start date as the publication of a Commission proposal and this takes no account of what may be a protracted period of drafting.

In the formulation of environmental policy, there is often relatively little engagement with other policy debates. There is a risk of environmental policy being confined within an essentially technocratic paradigm which looks for solutions through the prohibition or regulation of an activity or some technological fix. Environmental policy needs to be sensitive to social differentiation and other social phenomena. Thus, in the case of cars, which are a major source of pollution, producing cleaner cars or fuels is of itself not enough. One also has to understand and tackle the deeply rooted social attitudes and behaviour patterns which manifest themselves in a 'car culture'.

A third possible difference between environmental policy and other policy sectors is that it is more politicised in terms of public attention. The environment does have high salience as an issue among the European public, and cleaning up bathing beaches or making cars less polluting is one way that the EU can impact positively on the lives of its citizens. Successive Eurobarometer surveys showed that the percentage of respondents in the EU considering protecting the environment and fighting pollution 'an immediate and urgent problem' increased from 72 per cent in 1986 to 85 per cent in 1992, before falling back marginally to 82 per cent in 1995. However, when asked in 1995 whether environmental protection or economic development should be the priority, only 18 per cent suggested that the environment should be given a higher priority with 72 per cent selecting the option that 'Economic development must be ensured but the environment protected at the same time' (Eurobarometer, 1995, p. 15). These are by no means incompatible objectives, and some forms of environmental protection can create jobs (although often not in regions of high unemployment) but the pattern of response does suggest a certain superficiality in public concern about the environment. In many ways it is the non–governmental organisations that articulate the concerns of the 'interested public' that have a greater impact on decision-making and we shall return to their role in Chapter 2.

A fourth and possibly the most significant difference is in terms of the role of 'lead states' in the formation of environmental policy. If one takes the CAP as a contrast, each member state has its own agenda in terms of maximising the financial and other advantages it derives from the policy. Southern member states have a different agenda from those in the north because they are concerned with different commodities, but all the member states are interested in the outcome of the bargaining process. France and Germany play a pivotal role in the development of the CAP, but it is often a role of resisting or moderating change rather than leading the CAP in a new direction.

In environmental policy there are a few member states that have a strong commitment to the development of policy while others are indifferent or even regard it as an unwelcome imposition. Thus, 'the policy process in the environmental area is typically driven by a small number of member states which are significantly more environmentally progressive than the rest' (Sbragia, 1996, p. 236). The three leading states are Germany, Denmark and the Netherlands, although since the last enlargement one might add Sweden to the list. Liefferink and Andersen note (1998, p. 258), 'Among the new member states, concern about the possible consequences of membership for national environmental policy is probably greatest in Sweden.'

The position of these states is driven by their domestic politics. Indeed, it may even be the case that domestic environmental policies 'are increasingly designed with a deliberate view to the possible impact on EU policy-making' (Liefferink and Andersen, 1998, p. 255). 'Of the troika, Germany has been regarded as the key member state driving the progression of environmental policy' (Sbragia, 1996, p. 239). Germany's role remains important, but it has also become more ambiguous:

> Owing, among other things, to severe economic problems and an inability to catch up fully with the shift in EU environmental policy from a standard-oriented to a more processual approach, Germany has gradually lost the position of pusher-by-example that it had during the 1980s. What is left is not always clear. In some cases, such as climate policy, Germany appeared as the most reluctant 'green' member state. (Liefferink and Andersen, 1998, p. 268)

The establishment of a 'Red–Green' coalition government did not have an umambiguous effect on Germany's role in the environment Council. Although Germany had a radical Green (Jurgen Trittin) as environment minister, he did find himself in the embarrassing position of having to change his position in the environment Council following new instructions from his political masters at home. The German presidency in the first half of 1999 was not characterised by dramatic progress on environmental policy, with a number of dossiers left on the table without agreement, although admittedly the crisis in the Commission did not help.

Even when Germany proposes items for the environmental agenda or even suggests first drafts for legislative proposals, converting these proposals into law does require a broader coalition, particularly when there is resistance from elsewhere in the Commission. The core group of member states has to be careful not to be seen as a clique – a factor which inhibited the formation of a Danish-led 'Nordic coalition' following the admission of Finland and Sweden (Liefferink and Andersen, 1998, pp. 263–4).

It is important not to overstate the distinction between 'leader' and 'laggard' states. Although one can identify the lead states, it is less clear who the laggards are (for example, it is hard to categorise Belgium, France, Italy and Luxembourg). Britain might be seen as a laggard, but in practice it has often been the country with the most innovative ideas about policies to deal with the interface between agriculture and the environment. As Flynn notes (1998, p. 696), 'Britain has clearly evolved from the more minimalist and hostile stance of the early 1980s to emerge as a medium-positioned state in the league of environmental leaders and laggards.'

Even within the 'green core', 'Alliances between the "green" member states ... are by no means given. They have to be informed on an issue-by-issue basis and remain liable to defection' (Liefferink and Andersen, 1998, p. 262). These states have other concerns and national interests apart from environmental policy. There are limits as to how far they are prepared to go in their support for environmental policy. A Commission official commented to one of the authors, 'Even environmentally enlightened states say, "Commission, it's none of your business to tell us how to spend our money."'

One final point needs to be noted about EU environmental policy, although it does not differentiate it from other policy sectors. The makers of EU environmental policy lack one vital policy instrument – the ability to levy taxes. Pollution taxes have been increasingly emphasised as an effective instrument of environmental policy in recent years, but environmental measures that involve tax considerations are subject to an unanimity rule.

Commission services: the environment Directorate

When people talk of 'the Commission' in discussions of the European Communities, they often mean 'Commission services', the directorates-general and other units that make up the bureaucracy of the Commission, rather than the commissioners who constitute the college of commissioners. The focus here will what used to be known as DG XI, before the Prodi Commission decided to abolish the system of Roman numerals and revert to simple titles. An inherent structural and institutional problem is its relationship with other directorates-general, given the divergent policy interests of the different DGs, although in the case of the environment Directorate these policy conflicts have also been exacerbated by a number of factors discussed below.

Before looking at the difficulties of the environment Directorate we shall consider some of the weaknesses of the Commission system as a whole as revealed in the March 1999 report of the Committee of Independent Experts 'on allegations regarding fraud, mismanagement and nepotism in the European Commission', a report which led to the resignation of the whole Commission and its replacement by a largely new body under the presidency of the former Italian prime minister Romano Prodi.

Media comment understandably focussed on some of the more scandalous revelations, such as those about the Commission's Security Office. However, the report also drew attention to a number of worrying structural deficiencies in the operation of the college of Commissioners and Commission services and the relationship between them. The committee was concerned about the separation between the political responsibility of the Commissioners and the administrative responsibility of the director-general and the services. There was often a loss of control by the political authorities over the administration they were supposedly running. Directorates-general were compartmentalised, with as many fiefdoms as there were commissioners – a sorry prospect if one is trying to spread the impact of environmental policy across the work of the EU as a whole. The direct management responsibilities of the Commission had increased substantially since the 1990s, but this had not been matched by an emphasis on policy implementation.

This dismal picture in part reflects the fact that the European Commission has a number of special characteristics as a bureaucracy which could be captured in part by a number of alternative models. Page (1997) has characterised the main models as: the EU administration as a dynamic core of Europe; the EU administration as an international bureaucracy; the EU administration as a typical continental bureaucracy; and (following the American example of dispersed power) the EU administration as a subgovernmental bureaucracy. Page particularly emphasises the fragmentation of the EU bureaucracy. He acknowledges that all complex and diverse government bureaucracies exhibit some measure of fragmentation but 'there are two features of the EU bureaucracy which might be expected to increase [its] fragmentation' (Page, 1997, p. 18). The first is that the EU is a multinational institution where cultural differences, especially those embodied in language, may lead to a slower decision-making process and inhibit the exercise of managerial authority. A specific consequence in the area of environmental policy is the degree to which national models of environmental protection have been adopted as Community norms. Regulatory emulation of German policy is evident in the way in which the 'precautionary principle' appropriated the language and principles of German law in the late 1980s, as did the Large Combustion Plants Directive during the same period. It is no surprise to learn that Germany is 'substantially overrepresented' among officials in the environment Directorate (Page 1997, p. 58). The Netherlands is also important, having provided two directors-general in succession – Laurens-Jan Brinkhost (1986–94) and Marius Enthoven (1994–7).

The second point made by Page is that there is an absence of a single focus of authority, with most decisions reflecting a compromise between multiple European institutions, member states and interest groups. The fact that it has been possible to inject any dynamic at all into environmental policy is the result of the actions of a small group of states which have adopted a leadership role, a point which this book discusses further below in the section on the Council of Ministers.

'Like most national government ministries, the real action in policy-making takes place not at the level of the DG as a whole, but much lower down the organization' (Page, 1997, p. 31). Hull, who has served as an adviser to the Director-General of the then DG XI characterises the policy-making process in the following terms, although it may be that he underplays the extent to which the environment directorate relies on external expertise:

> The early thinking on any proposal takes place usually in the office of one Commission official who will have the responsibility for drafting legislation. The individual who is responsible for the initial preparation process . . . will find that when the final proposal is adopted by the Council it usually contains 80 per cent of his or her proposal. At the beginning he or she is a very lonely official with a blank piece of paper, wondering what to put on it. (Hull, 1993, p. 83)

What has been said so far could apply equally well to any DG in the Commission. However, part of the standard analysis of the environment Directorate is that it has been a relatively weak directorate-general relative to others in the Commission. This argument has been put particularly forcefully by Michelle Cini in her thought-provoking comparative analysis of the competition policy and environment directorates. Cini argues (1997, p. 81): 'DGXI is generally considered to be a weak DG within the Commission. Its inability to win arguments or to have its priorities translated into EU priorities provides ample evidence of its marginal character.'

The argument about the weakness of the environment Directorate advanced by Cini and others has a number of strands which need to be considered in turn. One central argument is that the environment Directorate has been affected by its comparatively late arrival on the political scene. Given that DG XI was established as the environment Directorate following the reallocation of the responsibilities of the former DG XI (internal market), this may seem a strange argument to make nearly 20 years later. Nevertheless, Page draws attention to the considerable stability in the structure of the Commission (his remarks do not take account of the changes made after spring 1995): 'The number of directorates general has grown by only three since 1972. Of the twenty-three directorates general, sixteen are identical or very similar in functional responsibilities to those created in 1972' (Page, 1997, p. 30).

The environment Directorate also has a very small staff compared to other directorates-general, particularly when one considers the range of responsibilities which are contained in its remit. Page's figures show (1997, pp. 32–3) that it ranks fifth from bottom among directorates-general in terms of its staff, 119 in 1993, although this figure cannot take account of the large number of temporary staff, allowing Weale (1996) to arrive at a figure of 450. Indeed, Page goes on to use a figure of 336 officials of all types in the environment directorate (Page, 1997, p. 63). Page makes the point that what was then DG XI was one of only two DGs in which over half the staff are non-establishment (the other was DG XXIII, Enterprise) (Page, 1997, p. 62). In contrast, only one-eighth of the officials in the agriculture directorate (formerly DG VI) are non-establishment and the largest category of these are seconded national officials, suggesting that member state governments think that it is important to get their officials inside the agriculture directorate. Whereas 47 per cent of DG VI non-establishment officials were seconded in 1992, the corresponding figure for DG XI was 11 per cent (seven). The environment Directorate's non-establishment officials are mainly 'external resource' officials – that is, they are brought in, under a variety of arrangements, to plug staffing gaps rather than being national civil servants seeking to get on the inside of a locus of power.

What is also significant is that DG XI had the smallest number of staff per unit (seven) of any directorate-general. The unit is the basic organisational level of the Commission at which, in many respects, the real work of drafting and monitoring is undertaken. This lack of staff in a relatively new policy area reflects the failure 'of any attempt by the Commission to assess in advance the volume of resources required when a new policy was discussed among Community institutions' (Committee of Independent Experts, 1999, p. 6). It could be argued that these low levels of staffing are misleading as the environment Directorate tends to make much more use of staff on short-term contracts and outside consultants than other directorates-general. This is not, however, a mark of merit. The 1999 report of the Committee of Independent Experts found that 'The use of outside assistance . . . demonstrates that the Commission has failed to tailor its human resources to its needs' (Committee of Independent Experts, 1999, p. 5).

Given its lack of funding and staff shortages, the environment Directorate also relies to quite a considerable extent on the expertise of various environmental groups in Brussels. Although the use of consultants and other external contacts does bring more outside inputs into the thinking of the DG than one might find in a long-established and rather insular directorate-general like agriculture, it is also a sign of organisational weakness. Contrary to popular myth, the EU bureaucracy is understaffed for the tasks it has to undertake, and this problem is particularly apparent in the environment Directorate.

Perhaps the most serious problem that the environment Directorate has faced is the reaction to it from other directorates-general. When it was first set up, committed environmentalists tended to be recruited to its ranks. 'One senior DG XI official has remarked that from the very start this gave the DG a reputation for being dominated by what he called "ecological freaks"' (Cini, 1997, p. 78). The reputation of the environment Directorate as being peopled by individuals who are environmentalists first, and Commission officials with wider interests second, does seem to have pervaded the Brussels administration. It has diminished to some extent as personnel have changed: by 1999 there was a British director-general with a pragmatic focus on achieving results rather than defending the purity of policy proposals. Nevertheless, such reputations tend to persist even when they have little basis in reality. 'In addition, its particularly technical focus and what seems to be a disregard for the political dimension of policy does not help to raise the profile of the DG from that of a minor league player' (Cini, 1997, p. 81).

Is this 'minor league' argument still valid? Cini refers to 'the perception that its policies seem far removed from mainstream Commission policies' (Cini, 1997, p. 81), yet successive treaties have strengthened the emphasis on environmental policies and the obligation of other parts of the Commission to take environmental priorities into account in their own policy-making. Brussels insiders point to the fact that the environment Directorate offices have been brought together in a sparkling new building with an impressive atrium, but one that is located a long metro ride from central Brussels. Weale (1996, p. 598) pays tribute to 'the administrative and diplomatic skills of the director-general throughout this period, Laurens-Jan Brinkhost'.

Anecdotal evidence nevertheless suggests that problems remain. One of the authors (Grant) was involved in a conference on an air pollution topic organised by DGs XI and XII (research) at which the DGs for energy and for transport were conspicuous by their lack of involvement. On another occasion, at a workshop in Brussels in 1998, a DG XI official expressed her frustration at the resistance of other DGs to a project she was working on. She commented on 'how difficult it is to work with another service. There are conflicts, problems and negotiations at every level.' Sbragia (1996, p. 245) sums up the problem well:

> Environmental proposals which come out of the Commission have at times been through a bruising internal battle, particularly those which are linked to the single market. In general, the DGs concerned with industrial and economic affairs will have tried to weaken or block proposals introduced by DG XI. The DGs concerned with economic affairs typically have far greater resources and technical expertise than does DG XI,

and the interest groups linked to such DGs have much greater status and influence and status than do the environmental groups which have access to DG XI.

Of course, all DGs may face opposition to their proposals when they go before the Commission meeting as a collegiate body. However, the environment Directorate is often less able to rely on the support of other directorates-general who in turn often have well-established alliances among themselves. Business interests have formed alliances with DG III (industry and internal market) and DG XV (or DG VI in the case of the farming lobby) to ensure that draft proposals from DG XI were blocked or weakened. This may change to some extent now that the old industry portfolio has disappeared as the result of the Prodi reforms. In its place is a new enterprise department that will also deal with small and medium-sized business and information technology.

Even when the environment Directorate has been able to form alliances with other directorates-general, these alliances have often been rather fragile and open to effective countervailing challenges from business-based coalitions. The driving force behind the proposal for a climate/energy tax was the environment and energy (DG XVII) DGs. 'This alliance was based on common interests representing EC climate policy, but their motivations varied. DG XI sought to promote environment interest, i.e. coping with the greenhouse problem, while DG XVII's major concern was to promote energy efficiency which would in turn promote energy security' (Skjaerseth, 1994, p. 27). In any case, the tax proposal met fierce opposition from DG XXI (taxation), backed up by the 'most business–"friendly" DGs responsible for economics and the Internal Market' (Skjaerseth, 1994, p. 28). Faced with this alliance, which was backed by Commission president Delors, the environment commissioner had to give in (see Chapter 5 on climate change).

Weale and Williams (1992, p. 59) note that 'those directorate-generals responsible for the development of the single market programme were at the traditional centre of the Commission's activities – finance and industry'. They suggest that one way out of this problem is to see environmental policies and economic policies as complementary with an 'ecological modernization' perspective seeing environmental protection as an essential precondition for growth and development. Business interests are prepared to subscribe at least in principle to this idea, but its impact on practice may be more limited. Thus Porta states (1998, p. 173), 'both governments and companies recognise today that environmental protection and economic development must go hand in hand . . . the problem is that this concept of Sustainable Development is not suitable for use as such in business practice'. Weale and Williams admit that the relationship between economic competitiveness and environmental protection and sustainable development remains ambivalent and ambiguous. While refusing to admit that the integration of environmental policy with other policies 'is a forlorn hope', they concede that 'the conditions for its success still need to be discovered' (Weale and Williams, 1992, p. 62).

One initiative to promote policy integration was made during the British Presidency in 1998 when it was suggested that the transport and environment directorates-general should be merged under a single commissioner. No doubt the British government had its own domestic merger of the transport and environment departments in mind when making this proposal. However, such an approach overlooks the rather different mission of the transport directorate-general, as reflected in the 1993 White Paper on Growth, Competitiveness and Employment 'where infrastructure, in the form of trans European networks, is used to promote the twin objectives of competitiveness and cohesion' (Vickerman, 1998, p. 131). Indeed, Britain's deputy prime minister, John Prescott, was obliged to admit that while 'It would be good to have a single commissioner for [transport and environment] . . . in itself this is no guarantee that problems will be solved, because this also involves industry and competition issues' (*Financial Times*, 28 April 1998). Nevertheless, the restructured Prodi Commission combined responsibility for transport and energy (along with relations with the Parliament) under Vice-President Loyola de Palacio. Whether this results in a more integrated and environmentally friendly strategy remains to be seen.

Although this discussion has focussed on structural issues, the personality and skills of the individual commissioner responsible for environment policy can make a difference to the policy-making process – although this impact is as likely to be negative as positive. In the context of agriculture policy, Grant has argued that 'It does matter who is the member of the Commission responsible for agriculture. An effective Commissioner can play a more central role in the decision making process' (Grant, 1997, p. 149). Since he became agriculture commissioner, Franz Fischler has steered agricultural policy in a more environmentally friendly direction against considerable resistance. For example, the needs of animals have become regarded as a legitimate policy consideration in a way that they never were before.

As Page (1997, p. 154) notes:

> A strong commissioner, or an astute one, can help secure the collective support of the College of Commissioners for a controversial initiative. For example, the support of Carlos Ripa Di Meana was crucial in gaining acceptance for DG XI's position on stricter car emission standards in 1989.

However, Di Meana's flamboyant style had its downside. He became embroiled in a series of arguments with the British government in which he threatened legal action over a wide range of issues, actions which attracted considerable publicity. However, these disputes 'transformed what might have been comparatively minor wranglings over the pace of implementation and scope for national interpretation of Directives into matters of high politics' (Lowe and Ward, 1998, p. 21). When the Commission threatened to stop work on the controversial Twyford Down road scheme, the prime minister, John Major, raised the stakes by threatening to block the proposed extension of majority voting for environmental legislation.

Moreover, Di Meana's departure in 1992 left something of a vacuum. Karel van Miert then served as acting commissioner from 1992 to 1993, following which Iannis Paleokrassas held a two-year tenure before Ritt Bjerregarrd followed with a four-year term from 1995 to 1999. As a consequence of these disruptions, 'DGXI's internal influence . . . waned. The Directorate appeared to lack strong political leadership or support as environmental attention declined across Europe' (Lowe and Ward, 1998a, p. 24). The environment Directorate has never been led by a commissioner of, for example, agricultural commissioners of the stature of Mansholt, MacSharry or Fischler. One reason may be because of the perception that 'There are only seven real jobs to be done within [the Commission's] ranks' (*The Economist*, 27 March 1999) and environment is not one of them. The environmental commissioner from 1995 to 1999, Ritt Bjerregaard, who had been education and social affairs minister in Denmark, faced some criticism:

> Publishing diaries critical of many of her fellow commissioners undermined her ability to get anything accepted by the EU executive, let alone governments. Combining a hectoring, school-mistressy tone with a frosty charm that has earned her the nickname 'Ice Maiden', she does not easily come over as a global wheeler-dealer. (*Financial Times*, 13 November 1997)

A strengthened cabinet did, however, help her to make progress in areas such as car recycling and air pollution and she prioritised three areas of activity: 'greening' the CAP; helping East European governments to meet environmental compliance standards; and better enforcement of existing EU environmental legislation. The first of these, however, is heavily dependent on cooperation from another DG, while the last depends on the cooperation of member state governments.

When the Prodi Commission was being assembled, Ms Bjerregaard's fate seemed to depend in part on whether she would be more of a political nuisance in Denmark or in Brussels, and whether she was needed to maintain the gender balance of the Commission. However, with five women secured for the Commission, 'there was no way Prodi was going to accept . . . Ritt Bjerregaard simply to keep her out of Danish domestic politics' (*European Voice*, 15–21 July 1999, p. 7). Prodi flew to Copenhagen to secure an alternative Danish nomination and the environment portfolio went to Sweden's Margot Wallström. Wallstrom had no particular background in environmental policy which would be a disqualification in a more established area like agriculture. 'She got the job because Sweden wanted it and nobody else wanted it as much' (*European Voice*, 15–21 July 1999, p. 6), which says something about its relative importance.

Nevertheless, Wallström was rated as one of the most impressive commissioners in the hearings held by the European Parliament, displaying considerable confidence in dealing with questions despite her inexperience in environmental policy. She emphasised, however, that she did not underestimate the difficulty of integrating environmental policy with other areas.

Individual commissioners can make a difference, but the environment Directorate has failed to attract politicians with the kind of credibility and stature that gives them weight within the college of Commissioners. This reflects the fact that such figures are not attracted to the environment portfolio because they are aware that relative weakness of the environment Directorate has deeper, structural causes. The EU is clear that it wants an environmental policy and it has reinforced that commitment in the Amsterdam Treaty. Environmental pollution is a transboundary problem, tackling it enhances the legitimacy of the EU with its own citizens and is also necessary to ensure that Europe plays its part in devising and implementing global solutions to problems such as climate change.

However, there are also factors which limit the impact of environmental policy. Public opinion on the issue is somewhat shallow and subject to cyclical fluctuations in response to economic factors. In the 1995 Eurobarometer survey on Europeans and the Environment, 82 per cent of respondents considered that protecting the environment was an urgent and immediate problem. However, only 18 per cent considered that concerns about the environment should be given a higher priority than economic development. Support for 'green' taxes falls from 73 per cent to 44 per cent when it is suggested that they could slow economic growth and if they were likely to lead to a serious slowdown in economic growth, only 13 per cent of the European population would support them (Eurobarometer, 1995).

A second limiting factor is that government support for environmental policy varies considerably between member states, reflecting their different domestic politics. This factor is considered further in the following section on the Council of Ministers. Third, and most important, business interests are able to argue that 'Over-regulation stifles growth, reduces competitiveness and costs Europe jobs' (Porta, 1998, p. 167). The 'greening' of Europe is favoured if the opportunity costs in the economy are not too high.

The Council of Ministers

The European Council of heads of government was given a formal status by the Maastricht Treaty on European Union which gave it the duty of providing the European Union with impetus and political guidance. The extent to which the Council becomes involved in environmental policy is affected by the priority given to the policy arena relative to other matters of concern. Although the Council will from time to time issue proclamations on the environment, as at the 1989 Dublin summit, or take action to implement action programmes, as at the 1992 Edinburgh summit, it is much less involved in genuine debate about environmental policy than it is about, say, agriculture, let alone questions of 'high' politics such as budget and enlargement. In practice, most of the decisions on environmental policy are taken by the Council of environmental ministers.

The Environment Council is not one of the leading Council of Ministers which meets frequently and takes far-reaching decisions which have considerable cross-sectoral importance (General Affairs and Ecofin), nor can it be compared with the leading sectoral council (Agriculture) which drives policy in its sector. It is one of a number of middle-ranking councils which meet four or so times a year to process business in their sectors. Equivalent councils are those dealing with transport, industry, research, social affairs and fisheries. As is the practice in many other councils, it is the practice for an 'informal' council in a pleasant location to be held early in a new presidency. These informal councils may be used to attempt to perform an agenda-defining function since it is during these meetings 'that the holder of the Presidency is most likely to be in a position to push for national priorities' (Wurzel, 1996, p. 281).

Within the European sectoral-level decision-making process, the Council of Ministers can be regarded as the principal decision-making body. Golub notes (1996, p. 336) that

> all the technical expertise, agenda-setting and influence of the Commission and Parliament come to nothing unless these two institutions have the determination to risk paralysing decision-making processes, because ultimately power resides with the

> Council – with a qualified majority in Council, to be more precise.

Golub's conclusions are based on a study of the Packaging Waste Directive, but that was an important directive and there is no reason to believe that it was atypical. 'Despite the obvious window of opportunity for a packaging waste directive, the Commission, and to an even greater extent DGXI, failed to secure most of its original primary objectives' (Golub, 1996, p. 325). The Directive was not a middle-ground compromise between the Commission and the Council, but 'represented the lowest common denominator because the Commission was forced to capitulate to a qualified majority in the Council' (Golub, 1996, pp. 325–6).

Meetings of the Council are prepared by the Committee of Permanent Representatives (Coreper) and by more specialist working groups. At Council meetings only 'B' agenda items are under discussion, 'A' items having been settled at the level of Coreper and only requiring formal approval in Council. A study of 43 environmental acts by the Council in 1993 and 1994 found that in 29 cases 'the environment working group was the primary negotiating forum' (Andersen and Rasmussen, 1998, p. 587). Coreper's role appears to be to filter and approve the recommendations of the working group before they go on to Council. This might reflect the fact that environmental policy is a relatively technical area and the input of experts from national capitals is therefore of particular importance. In any event, 'in environmental policy Coreper does not have the role of central problem-solver. Its relatively unobtrusive role seems to be provide negotiators at working group level with a certain amount of room to manoeuvre' (Skou Andersen and Rasmussen, 1998, p. 589).

Reference has already been made to the role of the 'green troika' of member states creating a 'push–pull' dynamic in European environmental policy, and in particular to the key role of Germany. Paradoxically, the introduction of qualified majority voting may have made it more difficult to bring along the 'laggard' states. Unless one can count on the support of Finland and Sweden to form a blocking minority, the leading states have lost their veto on proposals which they regard as minimalist or which actually threaten to undermine their own higher national standards. Golub notes (1996, p. 329) that, in spite of the changes brought about by the Single European Act,

> power ultimately lies with the national representatives in the Council of Ministers . . . the Commission and the Parliament were unable to dictate stringent environmental standards in the packaging directive because, in the end, their objectives were unpalatable to a qualified majority of Member States.

Nor did the entry of such 'greenish' states as Sweden, Finland and Austria give the expected boost to environmental policy-making (Lowe and Ward, 1998, p. 23), although they may have been able to give greater emphasis to the environmental dimension in debates in the Agriculture Council. Indeed, it is in this Council rather than the Environment Council that some of the most significant battles of the next few years will be fought. The role of the Environment Council has been diminished somewhat with the 'move away from the legislative factory approach of the 1980s' (Lowe and Ward, 1998, p. 24) towards a new emphasis on partnership with industry and flexible implementation focussed on the achievement of improved environmental outcomes rather than the passage of particular pieces of legislation.

Every six months, in accordance with a treaty-governed rotation, the chair of the Council of Ministers passes to a new member state. Apart from the formal role of chairing meetings, the member state occupying the presidency is responsible for managing the business of the Council, including negotiating deals to settle unresolved policy problems. As the Community's activities have grown, the workload and significance of the presidency has increased accordingly. The six-month presidency can give some scope to influence the direction and pace of Council business, but only to a limited extent, as an agenda is inherited and then passed on. Six-month presidencies mean that countries may start an initiative or complete one, but it is rarely long enough to see them through all their stages.

In a comparison of the British 1992 and German 1994 environmental council presidencies, Wurzel (1996) found little difference in the way in which they were run. The British used their presidency in 1998 to stage the first informal joint meeting of the transport and environment councils at which they pushed their agenda to secure a greater integration between the two policy spheres. The British also secured an agreement with their three successors as president of the Environment Council (Austria, Germany and Finland) to conduct an audit against common criteria and targets of the progress each country makes during its six-month presidency. The intention is to overcome the problem of short-term initiatives being launched by individual countries holding the presidency. What this leaves untouched is the issue of whether a six-month presidency is still appropriate or whether one might have one-year presidencies made up of a partnership between a larger and a smaller country (in practice, small member states rely on help from their neighbours during their presidencies, e.g., Luxembourg from other Benelux countries).

Andersen and Rasmussen (1998) compared the success rates of four presidencies in 1993 and 1994 in securing the passage of environmental acts and found that Germany and Denmark, both conventionally regarded as 'lead states', were the most successful, followed by Greece and then Belgium, which finalised only three items. They attribute this disappointing record to the absence of a federal environmental ministry in Belgium which 'means that the institutional background for setting and securing priorities in the Environmental Council are weak' (Andersen and Rasmussen, 1998, p. 592). Given that the record of Belgium in at least one other Council was disappointing (Grant, 1997, p. 174), it may be that domestic political tensions in the country also have an impact.

What is evident is that what usually happens in the Council of Ministers is a process of intergovernmental bargaining which 'typically . . . weakens the restrictions proposed by the Commission' (Sbragia, 1996, p. 247). The Council thus tends to lighten the shade of any 'greening' proposed by the Commission. Whether the Amsterdam Treaty's attempt to establish sustainable development as one of the main goals of the EU will make a difference to the decision-making process remains to be seen.

European Environment Agency

The primary objective of the European Environment Agency (EEA), which started its activities in November 1994 (although it was not fully operational until 1996), is 'to set up and manage a European information and observation network and ensure the dissemination of comparable information'. In its 1997 work programme it defines its goal as 'to be an independent source of environmental information, efficient and demand driven'. In general, the EEA does not seek to create more data but to make more effective use of existing data by aggregating it and standardising it. It seeks to provide 'knowledge in a nutshell' that is ready for use by policy-makers.

Based in Copenhagen, the EEA is an independent legal entity run by a management board which elects its own chair with one representative from each member state (plus Iceland and Norway), two members designated by the European Commission (one from DG XI) and two designated by the European Parliament. With a budget of 16.7 million euros in 1997, anticipated to grow to 25 million euros by the year 2000, it has a staff of around 60, of whom 30 are professionals. However, about half the budget is used for external services delivered by experts. The EEA also runs a data management network called Eionet which links it with the national environment agencies and various specialised centres and institutions which provide the national contribution to European-level data. This may sometimes involve aggregating data which has originated from industry sources.

According to its 1997 work programme:

> The mission of the EEA is to deliver timely, targeted, relevant and reliable information to policy making agents and the public to support the development and implementation of environmental policies in the Community and in the Member States.

Thus stated, its mission would appear to be a low key, technical one. However, there are three reasons why it could become more politicised. First, scientific information is no longer regarded as an objective statement of reality, but is seen to embody evaluative judgements, making it 'more contested and problematic'

(Wynne and Waterton, 1998, p. 126). Second, there is the risk of turf fights with DG XI, given that 'the boundary between "neutral" information and policy shaping is ambiguous and changing' (Wynne and Waterton, 1998, p. 126). In this connection, it is significant that 'The DG XI representative on the EEA management board has asserted a strict interpretation of [the establishing Regulation] to insist that the EEA's proper user relationships are limited to the Commission and the member state governments only' (Wynne and Waterton, 1998, p. 127). Third, although members states have sought to ensure that the flow of information to the EEA comes from 'official' sources, non-governmental organisations could use the reports published by the EEA as a basis for policy demands. As the EEA itself has stated (1995, p. 1):

> One good bathing water quality map . . . can speed progress towards better waste-water management in coastal resorts more effectively than a barrage of expensive technical studies, punitive regulations or combative protest campaigns.

The EEA was set up when environmental enthusiasm was at its peak at the end of 1980s, and 'green' MEPs were pressing for an equivalent of the US Environmental Protection Agency. Such an executive agency would not have fitted in with the EU structure, but Commission president Jacques Delors backed something more modest. The EEA's initial stance has been a cautious one, its activities limited by the resources available to it and the need to avoid potentially damaging turf fights. Nevertheless, 'As regulatory cultures change towards more information-dependent styles, the policy importance of an information agency is likely to grow' (Wynne and Waterton, 1998, p. 124).

There have been suggestions for extensions of the role of the EEA. The European Consultative Forum on the Environment and Sustainable Development has called for the broadening of its mandate into that of a 'European Sustainability Agency'. More specifically, it has been suggested that the EEA might contribute to the independent review of the environment performance of all directorates-general that the Commission has proposed (Klatte, 1999, p. 6).

European Parliament

The European Parliament (EP) consists of 626 MEPs drawn from the 15 member states. The Parliament's legislative role under the Justice and Home Affairs and Common Foreign and Security Policy 'pillars' of the EU is very limited, the only Parliamentary involvement being consultation by the Presidency, which asks the Parliament for its views on the main policies adopted. It is under the European Community 'pillar' which includes environmental policy that the influence of the EP has grown. Craig and de Burca (1998, p. 66) describe how the EP has gradually been transformed from a representative body with a consultative role into a directly elected chamber with an important role in Community environmental policy.

Building on its original right simply to be consulted before legislation is adopted, the Parliament's powers were first extended with the introduction of the two budgetary treaties in 1970 and 1975, then with the introduction of directly elected Members of the European Parliament (MEPs) in 1979. Legislative powers were introduced under the 'cooperation' procedures of the Single European Act in 1986 and then under the 'co-decision' procedure provided for by the 1992 Maastricht Treaty, expanded to include other policy areas, including environmental policy, under the Treaty of Amsterdam.

Under the old consultation procedure for adopting Community environmental measures, the Parliament had little real power and the Council was often criticised for acting without waiting for the Parliament to give its opinion (Shaw, 1993, p. 79). In 1986, the Single European Act introduced the cooperation procedure which added a second Parliamentary reading to the legislative process, but this procedure was not initially used for environmental policy. In 1992 the Maastricht Treaty extended use of the cooperation procedure to include Community environmental legislation, provided that it is proposed under Article 130s of the EC Treaty (the legal basis normally used for EC environmental law). However, although the cooperation procedure provided for a second Parliamentary reading on proposals for new environmental legislation, it did not give the Parliament significant powers to insist that its proposed amendments were incorporated into Community law. The Parliament rarely obtained what it ideally wanted – namely a specific response to each of the amendments it proposed on first reading (Shaw, 1993, p. 81).

Although, initially, environmental measures continued to be adopted under the cooperation procedure, the co-decision procedure introduced by the Maastricht Treaty allowed measures to be adopted jointly by the Council and the Parliament, with co-responsibility between the two institutions (Shaw, 1993, p. 82). If the Council and Parliament fail to agree, a Conciliation Committee adjudicates and a possible third reading may be added. If the Conciliation Committee fails to produce a compromise, the Council may still determine the final version of the legislation that it wishes to adopt, while the Parliament can only block the Council's preferred text. In practice, the Parliament has been reluctant to block legislation and this reluctance has been attributed to political pressure exerted on the Parliament not to cause a legislative blockage which could prevent any new measures being adopted at all (Shaw, 1993, p. 82).

In 1996, the Treaty of Amsterdam gave the European Parliament greater influence over environmental legislation by extending use of the co-decision procedure to replace the cooperation procedure for all environmental measures adopted with Article 130s of the EC Treaty as their legal base. Despite the limitations of co-decision, its use as the legislative procedure for environmental measures considerably strengthens the Parliament's role in the adoption of new environmental legislation. In most cases, the Parliament now has a significant role in determining what form Community environmental legislation will take. This will go some way towards addressing the criticism that a lack of democratic accountability is inherent in the EC environmental policy process (Bell, 1997, p. 62).

The European Parliament, and in particular its Environment Committee (first established in 1973), is generally seen as an institution that is concerned to actively promote an effective European environmental policy. Indeed, 'Much of the EP's success has been due to the ideological devotion and tireless efforts of Ken Collins and other members of its Environment Committee' (Golub, 1996, p. 326). However, as Golub points out, the Parliament does not necessarily follow the advice of its Environment

Committee. The parties of the centre-right in the Parliament, who became the largest single group after the 1999 elections, tend to be resistant to any calls for higher standards that impose significant costs on industry, while even the support of the Greens cannot be relied upon as they may prefer 'no action to bad action' (Golub, 1996, p. 327n).

The 'greening' effect of the Parliament cannot, thus, be relied upon. Indeed, the Parliament can be unpredictable because its involvement in an inter-institutional struggle for power may sometimes be given priority over policy concerns. Paradoxically, as it acquires greater powers, it will attract more attention from business interests and its 'green' stance may be diluted (Liefferink and Andersen, 1998, p. 266).

In plenary session MEPs sit according to the eight political groupings currently within the Parliament, the largest being the European People's Party (Christian Democrats), Socialists and the Liberal Democrats. The Greens are one of the smaller groups represented in the Parliament but they play a significant role not only in debates on environmental policy, but also by ensuring that the environmental impact of other Community policy measures receives careful consideration. However, the successive Green groupings within the Parliament since 1984 have been weakened by divisions over strategies and tactics; the dominance of the Germans, and the absence of other key countries such as France; and disputes about the desirability of European integration. 'Above all, Green MEPs have been faced with the unique green dilemma of attempting to represent alternative, decentralised policies within a bureaucratised, hierarchical structure' (Bomberg, 1998, p. 124). The Green Group in the EP which brought together the 28 members elected in 1989 displayed a willingness to compromise with allies in other parties, especially the Socialist group, turning the Green MEPs into 'credible parliamentary players' (Bomberg, 1996, p. 325). Green MEPs have often served as rapporteurs on issues that concern them. However, the post-1994 group was riven by national, ideological and individual divisions, displaying a charactersitic inability to 'form coherent positions on European unions beyond the broadest slogans' (Bomberg, 1996, p. 330).

Sbragia (1996, p. 246) suggests that the influence of the Parliament is greatest under particular sets of circumstances:

> The Parliament is most effective when it works with a Commissioner such as Ripa de Meana who is sympathetic to parliamentary power, knows how to use the Parliament's powers to extract concessions from the Commission, and who is very receptive to amendments proposed by the Parliament.

Golub's case study of the Packaging Directives showed that the Commission effectively abandoned the Environment Committee as the decision-making process worked through its different stages. Indeed, 'the Commission never endorsed the greenest amendment favoured by the Environment Committee' (Golub, 1996, p. 327). Although Golub emphasises that there have been several cases where the Parliament has altered or defeated the Council's common position, his overall conclusion is more pessimistic:

> all the technical expertise, agenda-setting and influence of the Commission and Parliament come to nothing unless these two institutions have the determination to risk paralysing the decision-making process, because ultimately power resides with the Council – with a qualified majority in Council, to be precise. (Golub, 1996, p. 336)

The Treaty of Amsterdam further strengthens the powers of the Parliament, although the poor turnout in the 1999 elections undermined its claim to popular legitimacy. There is a sense in which it is an institution for which great things are always just round the corner, but never quite seem to arrive. As a result, the 'greenest' institution in the European decision-making structure has been, in practice, one of the weakest.

European Court of Justice

The European Court of Justice (ECJ), the judicial branch of the Community, is generally considered separately from the three main 'political' institutions (the Commission, the Council and the Parliament). However, despite this formal distinction, it is widely recognised that the Court has played an important part alongside the other institutions in shaping Community policy (see, for example, Craig and de Burca, 1998, p. 78). The basic task of the ECJ is to ensure that Community law is observed (Shaw, 1993, p. 73). In this role, the ECJ has given substance to Treaty obligations and subsequent legislation on environmental policy by clarifying the meaning and status of Community law. The Court comprises 15 judges, assisted by eight or nine Advocates General, and gives judgements on cases involving the Community institutions, member states and on a wide range of issues referred to the ECJ by national courts. The Court also gives its opinions on the compatibility of international agreements, including environmental agreements, with the Treaty.

The Court has the power to impose financial penalties on member states which fail to comply with its judgements and the role of the Court in ensuring that Community environmental law is implemented and enforced in the member states is discussed in more detail in Chapter 3. What is important to note here is that the Court has played an important role in response to political imperatives (Shaw, 1993, p. 74) and has often given judgements that take Community environmental policy further than strict interpretation of the Treaty would initially have envisaged. Two cases illustrate this point and show the significance of the ECJ's role in strengthening and expanding the status of Community environmental policy.

In the *Danish Bottles* case (Court of Justice 1988) the Commission brought an action before the ECJ challenging the legality of Danish laws which required beer and soft drink containers to be returnable, arguing that this amounted to a form of disguised discrimination against foreign manufacturers and hence a barrier to free movement of goods under Article 30 of the EC Treaty (for a full account of this case, see Bell 1997, pp. 82–3). In its judgement, the Court set a new precedent for Community law, finding that it was permissible to use environmental protection as a reason for discriminating against foreign manufacturers. The Court went on to establish the general principle that environmental protection could be a legitimate barrier to free trade provided that the measure was proportionate to the end to be achieved, namely that the same level of environmental protection could not be achieved by any other measure which did not hinder free trade. The Court's decision on the *Danish Bottles* case was particularly important because it influenced attitudes towards environmental policy in the member states and the Commission. For member states, it raised the prospect that national environmental measures could be stricter than those in other parts of the Community, even if this amounted to a restriction of free movement in the single market. For the Commission, the case provided the justification for introducing new legislation that provided for stricter common environmental standards at a Community level in order to prevent distortions in the single market.

The Court went on to the extend the principles of the *Danish Bottles* case in the *Wallonian Waste* case (Court of Justice, 1992) which concerned a ban on waste exports by Wallonian authorities. The ECJ acknowledged that, since waste is a form of 'goods', any restriction on its free movement amounted to an infringement of the EC Treaty. However, the Court also held that restrictions on free movement could be justified on grounds of environmental protection because waste is a form of goods that has a 'special character'. The ECJ justified this by referring to Article 130r(2) of the EC Treaty which requires that pollution should be rectified at source. So, by giving waste a special status, the Court was able to avoid the otherwise logical conclusion that a ban on export of waste was contrary to the 'free movement' principles set down in the Treaty (Bell, 1997, p. 82, gives a more detailed description of this case).

The Court's decisions on *Danish Bottles* and *Wallonian Waste* considerably strengthened the status of environmental protection in Community law. This new status was formalised by the Maastricht Treaty in 1992 which expanded Article 130r(2) of the EC Treaty to allow Community legislation to include safeguards that permit member states to take strict provisional measures to protect the environment, subject to inspection of those measures by the Commission. So, although environmental policy is now well enshrined in the treaties, the role of the Court in strengthening and expanding its status should not be underestimated.

Conclusion

Article 3d of the Amsterdam Treaty replaces and strengthens Article 130r(2) of the Maastricht Treaty in terms of the obligation to integrate environmental requirements into all EU policies and actions:

> Environmental protection requirements must be integrated into the definition and implementation of Community policies and activities referred to in Article 3, in particular with a view towards promoting sustainable development.

The Commission sent a communication to the Cardiff summit in June 1998 suggesting how this might be done, initiating what came to be known as the 'Cardiff process'. In doing so, they frankly addressed one of the fundamental dilemmas of contemporary environmental policy:

> The progress we are making by classic environmental regulation will not be sufficient on its own. Most of our environmental problems have their origins in current practices in sectors such as agriculture, transport, energy and industry and we must look to these areas for their solution. Attention must also be given to our unsustainable consumption patterns. (European Commission, 1998, p. 4)

The communication puts forward some sensible if modest suggestions, such as that all key policy initiatives where an important environmental effect is expected should be accompanied by an environmental assessment and that the European Council should periodically review environmental integration into key sectoral policies. However, it is admitted that 'The implementation of this approach requires a strong commitment from all Community institutions' (European Commission, 1998f, p. 7). Will one be forthcoming? The Commission states, 'Fully integrating environment into other policy areas is a long term challenge requiring a step by step approach which builds on experience. This will eventually require consideration of its implications in all policy areas' (European Commission, 1998f, p. 8). Some of the key phrases here are 'long term challenge', 'step by step' and 'eventually'. Decoded, what this statement means is 'Don't expect too much too soon.'

The fundamental problem is that the strategy outlined by the Commission represents 'a break with our traditional sectoral decision making' (European Commission, 1998f, p. 4). Each policy sector in the Community is carefully demarcated and each has its own set of clients and its own particular policy dynamic. Thus, for example, simply put, the CAP is a device for transferring money from taxpayers to successful farmers and protecting them from international competition. The policy-making process in agriculture is not suddenly going to change because of a treaty requirement about integrating environmental policy. The Commission places its faith in 'high politics' overcoming 'low politics', arguing that a cross-sectoral approach 'can only come about by Heads of State and Governments assuming responsibility' (European Commission, 1998f, p. 4). But what are the incentives for them to prioritise the environment over other pressing policy issues of greater immediate concern to their domestic electorates?

When the Prodi Commission set up five special groups of commissioners to coordinate work in areas of interest to more than one commissioner, one was concerned with economic growth, competitiveness and employment; one with a social policy issue (equal opportunities); two with institutional issues (internal reform and inter-institutional relations); and one with external relations. Despite draft rules stating that these groups are designed to ensure better coherence within policy areas, no group was set up to coordinate work on the 'Cardiff process', suggesting that a relatively low priority was attached to it.

The environment has often taken second place to economic goals. The economic focus was clear in the Treaty of Rome, establishing the European Community in 1957, which had as its core objective completion of the internal market ensuring free movement for goods, services, labour and capital between the member states. These priorities persisted even after an environmental policy had been adopted:

> One of the most striking examples of non-integration is – without any doubt – the decision in 1985, to proceed with the completion of the Internal Market. Neither the Commission's White Paper on which this decision was based, nor the Cecchini report on economic advantages of 1992 did in any way consider the likely environmental consequences of the initiative. (Klatte, 1999, p. 2)

In terms of institutional arrangements and decision-making processes, environmental policy has been something added to an existing set of arrangements. Consequently, it has had to conform to established rules of the game. Hence, despite a series of significant treaty reinforcements, environmental policy has often been marginalised within the EU. The EEA's report published as a contribution to the review of the Fifth Environmental Action Programme noted that the EU was 'making progress in reducing certain pressures on the environment' but 'This progress is not enough to improve the general quality of the environment and even less to progress towards sustainability' (Klatte, 1999, p. 3). There is a recurrent tension between the general growth-oriented objectives of the EU, reflected in the single market project and the euro, and the commitment to protect the environment. Citizen movements supporting the environment have grown in importance, but are often countervailed by well-organised business interests. It is to this battle for influence that we turn in Chapter 2.

2
Business and Environmental Interests

In both the formulation and implementation of policy, the institutions of the EU rely upon the respective expertise, public access and 'watchdog' capacity of NGOs and business groups. Each is central to EU goals in areas of economic and monetary union and environmental policy, such that their support is critical to effective policy outcomes. The first part of this chapter discusses the importance of business interests to the effectiveness of policy. The latter sections address the growing importance of the environmental movement in shaping the agenda of EU environmental policy.

There is no single business interest in relation to environmental policy which leads to demands for minimal regulation and the lowest possible standards of environmental protection. The impact of environmental policy differs by company, firm size and sector. Indeed, within a company, there may be a divergence of interest in relation to a particular environmental policy proposal. Coen (1998, p. 9) notes:

> BP is a large producer of both ethanol and fossil fuels and each has a separate profit centre. However, when faced with the possibility of the EU environmental legislation that gave tax concessions to synthetic fuel, the firm had to decide whether to lobby for fossil fuels or environmentally friendlier ethanol base products. In this case, the planners favoured the traditional extraction sector as they had just negotiated a new exploration programme.

It is important to note that there are a number of reasons why a particular environmental policy initiative may benefit a firm. There are benefits to be gained in terms of cost reduction by, for example, using energy more efficiently. A new regulation may create a new or larger market for environmental protection equipment that the firm produces. For example, in its 1997 annual report, ABB, a leading multinational company in power generation equipment, notes:

> The climate change agreement reached at . . . Kyoto . . . could influence many of ABB's markets. By pushing industrialised countries to reduce greenhouse gas emissions, and urging developing countries to continue infrastructure development without compromising environmental standards, the agreement promises to increase demand for ecoefficient technologies – one of ABB's strengths. (ABB, 1997, pp. 7–8)

Environmental regulations may serve the interests of a firm or sector by promoting market closure. They may raise entry barriers in an industry, or force less efficient or more marginal firms out of business, thus reducing competition. In particular, they may undermine firms which are seen to offer cut price competition and damage an industry's image by not adhering to 'quality' standards. Environmental standards may also act as non-tariff barriers which reduce competition from imports.

'Less tangible than profits, [another] reason to be green is enhanced reputation, better image, good publicity and customer loyalty' (Eden, 1996, p. 10). For example, the British supermarket chain, Tesco, deliberately portrayed itself as 'greener' as part of a strategy to reposition itself and move 'up market' (Higham, 1990, p. 17). Establishing such a 'green' reputation may assist firms to respond to unanticipated environmental regulations more effectively without sustaining permanent damage to the firm and its brands.

Nevertheless, firms generally have reservations about the 'imposition' of environmental standards by public authorities. It was such concerns that prompted the EU to establish a group of businessmen and civil servants known as the Molitor Committee to examine whether EU and national legislation was imposing unnecessary burdens on companies (Grant, 1998, pp. 160–1). The UNICE Regulatory Report, based on a survey of 2100 companies, found that, after tax and employment regulations, environmental regulations were selected by 57 per cent of companies contacted as having the greatest adverse effect on their competitiveness. In particular, they complained about environmental regulations. These were seen as not proportionate to hazard and risk and as too complex and prescriptive. Furthermore, it was felt that enforcement was inconsistent between countries; that technical definitions were poor or inconsistent; and that regulations were too difficult to understand (Porta, 1998, p. 167). 'The main impacts upon firms are that regulations lead to increased operating costs, increased capital expenditure, and diversion of management time' (Porta, 1998, p. 168).

These impacts are greater in small businesses because they are less likely to have specialist staff to deal with environmental questions and their administrative capacity is more stretched (Grant, 1998). UNICE (1995, p. 29) has argued that the 'diversion of scarce managerial resources may prove to be the most damaging of all of the many effects of regulations upon the competitiveness of SMEs'. Given that trade associations are usually dominated by larger firms, smaller firms may have fewer opportunities to influence the development of regulatory policy. However, smaller firms may be in a more advantageous position when it comes to compliance and enforcement. In their analysis of the European waste disposal industry, Brusco, Bertossi and Cottica (1996, p. 140) suggest that small firms and taxpayers may collude to tolerate lax enforcement. They conclude that 'small firms have captured not regulation, but enforcement and control'.

Some sectors of the economy, such as chemicals, pulp and paper manufacture, electricity generation, necessarily have a high profile in relation to environmental questions. However, there are very few sectors of the economy which are untouched today by environmental considerations. For example, the retail sector is obliged to consider the environmental impact of the transport of its goods by road and the waste implications of the packaging of the goods it sells.

Business influence on European environmental policy

The general argument advanced here is that business influence on European environmental policy is substantial and is present through the various stages of the policy-making process – with important implications for the overall effectiveness of policy. For the purposes of this discussion, business is equated with the leading multinational companies that are the principal players in European-level business organisations.

In some cases, business may be able to prevent action being taken on a particular proposal. The proposals for an EU carbon tax provide the clearest example of the direct effects of corporate lobbying in vetoing the emergence of a policy designed to limit greenhouse gas emissions' (Newell and Paterson, 1998, p. 685). The arguments put forward by industry came to form a crucial part of the framework of reference used by decision-makers. Lobbyists were able to erode the consensus among states within the EU in favour of a binding directive (Newell and Paterson, 1998, p. 686).

Even if a directive is passed, business lobbies may be able to undermine policies by refusing to cooperate at the implementation stage. In the case of the carbon tax, a number of member state governments were warned by their energy sectors that they would not meet voluntary energy efficient targets if governments proceeded with the tax proposal or imposed it unilaterally:

> Trade associations of appliance makers also obstructed European efforts to impose mandatory standards for energy efficiency on their products, by refusing to supply data and cooperate in an energy efficiency study upon which to base policy, delaying the efforts of the EU to meet its obligations under the Climate Convention. (Newell, 1997a, p. 205)

In exerting influence on the environmental policy-making process, businesses possess a number of advantages over other groups. First, business groups are the predominant category of European interest group, constituting 63 per cent of all European-level interest groups in one survey (Aspinall and Greenwood, 1998, p. 3). This numerical supremacy is reflected in a comparable level of activity.

A study of the 1994 packaging waste directive found 'that of the 279 lobbying entities that contacted DG XI as the Directive moved to adoption in 1990–93, just over 70 per cent represented trade and industrial interests, while less than 4 per cent represented environmental interests' (McCormick, 1998, p. 199).

Second, a multinational company will have several different types of organisation at the European level through which it can represent its views (leaving aside the possibility of persuading member state governments to adopt a position favourable to its interests). These range from product-level associations to informal dining clubs. This means that a message can be repeated in different locations or fora, or it can be subtly varied to reflect the complexity of the company's interests, or the stage of the decision-making process that a particular measure has reached.

The development by companies of their own European-level representation through increasingly sophisticated government relations or public affairs divisions has been documented over the years (see, among others, Grant, 1981; Grant, Paterson and Whitston, 1988; Cawson, 1997). One of the most detailed and careful empirical studies of this phenomenon at the European level has been carried out by Coen (1997, 1998). When government affairs directors of Europe's largest companies were asked to rank alternative channels in terms of their effectiveness in influencing policy issues, environmental policy was one of two policy areas out of seven (the other was trade) where European channels of action were seen as particularly important. An illustration of the kind of activity a multinational company may initiate was a 'stakeholder roundtable' organised by the American chemicals multinational Monsanto in April 1998. This brought together representatives of farmers' organisations and environmentalists to 'help identify attitudes in Europe to the introduction of genetically modified foodstuffs, and make a contribution to the development of an appropriate strategy within Monsanto' (SustainAbility, 1997, p. 3).

Some large firms have come together in informal dining groups. One of the longer standing ones is the Ravenstein Group, 'an elite dining club of government affairs directors operating from Brussels' (Aspinall and Greenwood, 1998, p. 13). Such meetings provide an opportunity to exchange inside information about policy developments and to review strategies and tactics. Another group is European Business Agenda, an informal alliance of British firms which originated with a dinner. Its compact nature means 'that group members can exchange gossip freely about developments in Brussels, and work very quickly together as issues demand' (Aspinall and Greenwood, 1998, p. 13).

Firms also operate through a range of associations at product, sector and cross-sectoral levels. At the peak of this pyramid of influence is the so-called 'business troika' of the European Round Table, the Union of Industrial and Employers' Confederations of Europe (UNICE), and the EU Committee of the American Chamber of Commerce. The European Enterprise Group, a group of government affairs representatives of leading European multinationals, acts 'as an intermediary body among the three' (Green Cowles, 1997, p. 130). The European Roundtable of Industrialists, made up of chief executive officers from 45 leading European companies, 'is arguably the most influential interest group in Brussels' (Green Cowles, 1998, p. 108). The EU Committee of the American Chamber of Commerce, representing the interests of American multinationals, is also highly regarded by the Commission. UNICE has been subject to criticism from time to time for being the least effective and efficient organisation in the troika. It underwent reforms in 1990, but these criticisms were renewed in 1998 with complaints about the range of issues it attempts to tackle, its reactive character and its consensual decision-making procedures (*Financial Times*, 25 August 1998). However, given the emphasis of the Round Table on selected 'strategic' interests, and the special interests of the EU Committee, it is UNICE that 'remains the workhorse of the business groups' (Green Cowles, 1997, p. 132). It is therefore the organisation that is most likely to be involved in a cross-sectoral business response to environmental issues.

Many of these responses will, however, come at sectoral level from organisations like the European Council of Chemical Industry Federations (CEFIC). CEFIC is based on the 'three pillars' of leading multinational companies, member state associations and bodies serving product sectors. CEFIC sets long-term objectives and tries to prioritise the issues it focuses on. Environmental protection is one of four strategic priority areas. 'It is important to ensure that health and environmental regulations are founded on a sound scientific basis and that their potential economic impact, including the international competitiveness of the industry, is fully taken into account' (CEFIC, 1998). Interestingly, however, CEFIC also considers that over the decades the focus of attention on the chemical industry has shifted from safety issues to environment and more recently increasingly to health.

A network of subsector and product associations is linked to CEFIC – either as affiliates or as sector groups of the main organisation. For example, the European Council for Plasticisers and Intermediates (ECPI) represents 28 companies making plasticisers or the raw materials used to produce them. Plasticisers are used in the manufacture of flexible plastic products such as polyvinyl chloride (PVC) used, for example, to make drainpipes. 'In recent years [the sector's products] have come under increasing scrutiny, even though the vast amount of scientific research demonstrates that their manufacture and use poses no significant risk to health or the environment' (ECPI, 1998). In a highly technical area, the ECPI commissions its own research, but also seeks to ensure that any results are presented 'in a concise and easy to understand format' (ECPI, 1998). Constant contact is made with customer companies who could be mobilised if their raw material was threatened. At the core of the ECPI's strategy is the deployment of its accumulated technical expertise to develop relations with decision-makers: 'Based on its extensive knowledge of the industry, many decades of experience, and its wealth of scientific data, the ECPI secretariat is also able to provide valuable input to legislative and regulatory authorities' (ECPI, 1998). In each product sector of the chemical industry, there are similar well-resourced associations deploying sophisticated political strategies. In common with all business associations, they are able to make an authoritative response to the often detailed technical content of DG XI proposals.

CEFIC itself can draw on a secretariat of eighty and 4000 specialist personnel sent by firms to contribute to expert committees. The specialist organisation representing the oil industry has 13 staff dedicated to environmental issues alone (Greenwood, 1997, p. 184). The leading seven environmental groups in Brussels had a staff of 36.5 in 1996 (Greenwood, 1997, p. 185). Staff numbers or committee experts are not everything. However, the Commission is significantly dependent on outside expertise in technical areas and this is often provided by business groups. A Commission official commented, 'Supposing that I wanted to find something out about biotechnology. I could ring CEFIC and they would have the leading expert in Europe at my desk by the end of the week. I would not bother ringing an environmental group.'

Business organisations also have contacts across a much wider range of directorates-general in the Commission than environmental groups. In one survey of environmental groups, 'only 16 per cent of groups had regularised contacts with more than two Directorates' (Lowe and Ward, 1998b, p. 101). As a consequence, 'It would appear that most environmental groups are still heavily dependent on DGXI for access and that the environmental lobby is still ghettoised within the Commission' (Lowe and Ward, 1998b, p. 101). In contrast, the chemical industry has contacts with directorates-general right across the Commission, including DGXI itself (Grant, Paterson and Whitston, 1988; Paterson, 1992). Business organisations often have their best contacts with the most influential DGs, ensuring that the influence they exert brings greater results.

Business can also draw upon contacts at the highest political levels when this is required. Environmental groups may have significant influence in setting the agenda and influencing new proposals, but when the point of decision is reached, large multinational companies and their organisations have access to members of the Commission and ministers and heads of government in member states. Coen forecasts (1997, p. 99) that 'the trend will continue towards increasing partnership between firms and the Commission at the European level'.

There is plenty of evidence of the structural advantages enjoyed by business interests arising from the congruence between their objectives and those of the EU. Green Cowles points out that there has been an increasing practice of chief executive officers of leading companies appearing with Commission officials to publicly endorse EU policies. She claims that the CEOs have acted as 'legitimizers for Commission officials who, as appointed office-holders, held no direct political legitimacy of their own' (Green Cowles, 1997, p. 130). Coen (1997, p. 99) also refers to a process of' mutual recognition and legitimization' by the Commission and large firms which 'has witnessed little adverse reaction from the general public'.

This special relationship derives from the fact that for decision-makers, business interests have greater legitimacy than environmentalists. Business values have come to play a more central role in a number of European countries, partly because of the influence of globalisation upon the economic priorities of governments. For example, business success is lauded by left-leaning governments, such as the Blair administration in Britain. Business can claim that only it can deliver the still widely held goals of economic growth and prosperity which are central to the project of European economic integration.

Environmental groups and EU environmental policy

The challenge of EU lobbying

Whilst there has been an expansion in environmental group activity at EU level, there are still relatively few environmental groups that maintain a permanent presence in Brussels. Instead, groups often pool their resources as the case of the self-proclaimed 'G8' (the eight largest environmental umbrella groups operating in Brussels: Climate Network Europe (CNE)); European Federation for Transport and Environment; Friends of the Earth Europe; Greenpeace International European Unit; WWF International European Unit; IUCN; and Birdlife International) and the EEB (European Environment Bureau) bears out. The EEB has consultative status with the Council of Europe and it represents 130 member organisations in 24 countries.

Even the largest international environmental groups, such as Greenpeace, Friends of the Earth and WWF (World Wide Fund for Nature), whilst maintaining branches in Brussels to coordinate their campaigns, are also often part of single-issue coalitions which attempt to represent a range of environmental groups in a particular issue area. Climate Network Europe (CNE) is an example of such a coalition.

As Grande (1996, p. 321) notes, 'The European system of interest mediation has its own specific features'. The European model of a multinational, neo-federal and open decision-making process presents environmental NGOs (ENGOs) with a number of distinct challenges (Mazey and Richardson, 1992). This system creates agenda-setting possibilities, but also structural weaknesses. Mazey and Richardson (1992) argue that the strengths include an ability to build Europe-wide coalitions of interests. Many Europe-wide organisations have meetings every four to six weeks in order to exchange information and ideas (Mazey and Richardson, 1992). Added to this is a sense in which the NGOs organised at Brussels level (albeit at different speeds and with different emphasis) are pushing in the same direction from a common platform. In contrast, it can be argued that many industries are in competition with one another, even where they are gathered loosely under the umbrella of a federation. In other words, there are perhaps fewer intra-interest rivalries among ENGOs. A further factor which may enhance the influence of ENGOs is their perceived ability (by the Commission at least) to contribute to the process of European integration by pushing for the extension of environmental competence at Brussels level. Most importantly of all, is the fact that NGOs are perceived to help the Commission to do its job more effectively. The small size of the Commission services means that it is very dependent upon outside sources for information and expertise which NGOs can provide. The size of the Commission also limits its oversight in matters of implementation; a role which ENGOs are in a position to supplement. Having the resources to both devise and implement practical solutions is 'particularly important in certain areas of the EU where local administrations may not be the best agents for service delivery' (Mazey and Richardson, 1992, p. 123). In general, environmental groups benefit from the fact that there is a tradition of close relations between DGXI and ENGOs founded on mutual support. Mazey and Richardson note (1992, p. 121) 'without NGO support DGXI might have died in its early years'. The environment Directorate is therefore keen to cultivate networks of support to bolster its position within the Commission power structure.

Too heavy a dependence upon DGXI is simultaneously, however, a structural weakness of environmental organisations in Brussels. The Commission provides direct and indirect (via contracts) financial support to ENGOs. The EEB, founded because DGXI needed an NGO movement as a counterweight to industry lobbies (Mazey and Richardson, 1992), is funded by the Commission to hold seminars and roundtables. This may compromise its position as a critic of EU environmental policy. The provision of Commission funding for the EEB is thought to shape its less confrontational lobbying approach (McCormick, 1998) such that some ENGOs are said to have been 'tamed' by this financial dependence (Mazey and Richardson, 1992, p. 122). Subsidies to the groups undermine the critical potential of the bureau vis-à-vis the Commission (Rucht, 1993). Its status under Belgian law means it must avoid overt political stances. Rapid decision-making is hindered by the fact that it has to reach a compromise between the differing tactics of its member organisations and the different national political styles of lobbying (ibid.). In 1992 DGXI spent roughly 6.5 million ECU on non-governmental environmental groups. About 10 per cent of this funding is set aside for the running costs of these groups (core-funding). As a result, up to half the group's annual budget comes from the EU (Rucht 1993). In 1990 ECU50 million was also made available to environmental groups for projects and research aimed at protecting the environment (Marks and McAdam 1996, p. 115).

The nature of the relationship has attracted fire from other quarters too. Industry, in particular, has been critical of what they consider to be 'agency capture' of the environment Directorate by environmental groups. 'Capture' or not, the organisational weakness of DGXI means that NGO's influence does not stretch as far as their industry counterparts who enjoy strong relations with the more powerful DGIII. Because of this, industry groups are in a stronger overall position within the policy-making process. It is also possible to question the extent to which 'agency capture' really occurs at EU level (Grande 1996, p. 322). As Mazey and Richardson argue, 'In practice, it is virtually impossible for any single interest . . . to secure exclusive access to the relevant officials, let alone to secure exclusive influence' (1993, p. 10).

A further structural weakness of environmental groups is the lack of resources which would enable them to participate in the policy process from initiation though to implementation. Keeping track of policy initiatives is a major undertaking for groups, made easier by the pooling of resources within Europe-wide networks.

Policy with a significant bearing on the environment emerges simultaneously from a number of directorates-general, creating a monitoring problem of huge proportions for often poorly financed ENGOs. Only the better resourced NGOs are able to orchestrate their activities such that they are in a position to lobby a number of DGs simultaneously and oversee the development of policy across a range of areas of relevance to the environment. WWF, for example, have lobbied the EU on its development cooperation policy both with third parties (such as ACP (African-Caribbean countries) and ALA (Asian and Latin American countries)) and within the EU, channelled through the regional and cohesion funds and the PHARE programme for Central and Eastern Europe. The group have also lobbied for reform of programmes addressed by powerful DGs which have an enormous environmental impact such as the CAP (Mazey and Richardson, 1992). The process of following the life-course of a policy from initial development through to implementation and maintaining good contacts with influential actors throughout is further complicated by the high turnover of Commission staff and the volume of short-term part-time staff (Mazey and Richardson, 1992).

The influence of environmental groups upon the direction and content of EU environmental law is also, of course, relative. The increasingly effective representation at Brussels level of interests counter to their own aims severely curtails the impact of the lobbying efforts of 'green' groups. The fear that the presence of powerful industry groups will undermine the power of ENGOs is fanned by the simultaneous pressures upon DGXI to work closely with industrial interests and upon the environmental unit of DGIII to play a greater part in the development of EU environmental policy (Wilkinson, 1997). In addition, the resources that industry groups provide the Commission with are more important for its purposes. As Rucht (1993, p. 89) notes:

> If environmental groups refused to cooperate with the Commission, this would not really pose a threat to the credibility of the rules and regulations it must prepare; but this would not be the case where the Commission depends upon the input of expertise from industry lobbyists.

The wider point is that industry lobbies are often better resourced with personnel to provide the Commission with up-to-the-minute responses to its requests for information. The industry federation CEFIC has a staff of 140 compared with the European Environment Bureau whose full-time staff numbers three (Rucht, 1993, p. 83). The other comparative point about industry federations is that their goals are more easily achieved than those of ENGOs. In many cases they are arguing for no change, which is often a more attractive and convenient option than pushing through a controversial proposal against bureaucratic resistance. In other words, vetoing or watering-down a proposal is more easily achieved than securing its successful passage through the policy-making process.

Nevertheless, lobbying the institutions of the EU serves a number of functions for environmental groups. It becomes especially important when national governments adopt more recalcitrant positions. Therefore, recourse to the institutions of the EU is often for strategic reasons. It serves short-term goals that cannot be met by national-level lobbying. Mazey and Richardson note (1992, pp. 116–17) 'In the environmental sector, groups at the national level are often in conflict with their own national administrations and hence see the EU as an alternative arena in which to exercise influence'. Grant notes, for example, 'the case against Britain's "dirty" drinking water was fuelled by information from environmental interest groups' (1993a, p. 28). The resort to the use of the European Court of Justice (ECJ) by groups such as Friends of the Earth in the face of a 'laggard' and Europhobic Thatcherite government in the UK is one manifestation of this trend. Sometimes NGOs can play the Commission off against their national government and vice versa in order to extract the greatest gain in terms of environmental policy reform. This process can jeopardise close relationships with national government departments nurtured over many years, but it can also bring about effective results. In the drinking water case above, relations with the government were not strong, so the risk posed by bringing Brussels on side was one worth taking.

Despite this, Lauber argues (1994, p. 257), 'The EU as a political structure offers substantially fewer opportunities for the environmental movement than the more advanced of its member states.'

Environmental organisations have no right to information, no right to be consulted in advance of a measure taken by EU organs and no right of standing before the European Court of Justice. They can only file complaints according to Article 230 (formerly Article 173) and Article 232 (formerly Article 175) of the Treaty. Environmental groups are, of course, consulted, but only on an informal basis and when the Commission seeks information or support for specific projects. Often consultation is sought on proposals which are effectively fait accompli and where a seal of approval rather than meaningful input and direction is sought. In this respect, environmental groups are denied the same privileges as business, labour and consumer groups, reflecting both the goals of the EU and the low profile of environmental groups at the time of the drafting of the Treaty in the late 1950s (Lauber, 1994) compared to stakeholders central to the process of European integration. Changes such as the directive on the freedom of access to information on the environment which came into effect on 1 January 1993 and the establishment of the European Environment Agency, said to 'create an important point of access for the environmental lobbies to EU institutions' (Lauber, 1994, p. 258), may make it easier to participate meaningfully in EU policy-making on the environment in the future.

Nevertheless, the fact remains that 'In contrast to groups in various other policy areas, environmental groups have no formal rights regarding the policy process in the EC' (Rucht, 1993, p. 88). The process is ad hoc and arbitrary so that who gets what information and who is invited to which committee meeting are often political decisions made by Commission officials. Contact between environmental groups and the Commission is often on the basis of informal contacts. Rucht (1993) describes a reactive process whereby NGOs respond to Commission proposals having little choice but to adopt the agenda and gain access to information as quickly as possible. The Commission is also relatively insulated from public opinion, such that the need for the sort of public approval of EU action which NGOs are able to mobilise is less pressing. This is not to suggest that ENGOs do not have an impact upon the Commission's agenda. National ENGOs in 'lead' states from the green 'troika' are in a position to set the

Commission's agenda. Evidence of this is to be found in the promotion of principles like the 'precautionary principle' within EU policy debates, which has been used in German law since the late 1980s. The EU large Combustion Plant Directive and the Packaging Directive also emulate German policy regulatory patterns which ENGOs lobbied for at the national level. We see, through this process, the diffusion from the national to the EU level of the indirect influence of strong environmental lobbies from powerful member states (Sbragia, 1996). Groups can also use their government's position as the Presidency to influence the work plan and priorities of the EU as a whole.

For the most part, however, it is very difficult for environmental groups to find out early enough in the policy process which proposals are being prepared by which section of a Commission department. Relying on formal consultations is rarely enough and most lobbying is done on an informal basis at 'social' gatherings and conferences. To be effective, groups have to lobby simultaneously at the national and European levels. This is where umbrella groups come in. They can orchestrate a Europe-wide campaign of activities with their member groups lobbying in the national capitals of 'laggard' states who may be obstructing the passage of an item of EU legislation. It is also necessary to put pressure on all institutions on a simultaneous basis, which is difficult for poorly resourced organisations. Despite claims by the Commission that a proposal has left its hands, it often continues to perform an ongoing role in the policy process (in relation to amendments put forward by the Parliament, for example) and needs to be continually lobbied.

The European Parliament is thought to be more sensitive to the interests of ENGOs and therefore provides a key point of contact in the policy process. Often the best way for ENGOs to influence the Parliament is to draft text in the form of a resolution, amendment or question. This involves lobbying the rapporteurs of the committees asked for an opinion by the Commission. Failing this, NGOs then try to find an MEP willing to put down an amendment on their behalf. It is difficult to mobilise sufficient support for a proposal, however, given that to stand a good chance of being accepted, amendments need the support of both the Socialist and European Peoples Party groups (the two largest groupings). The parliamentary intergroups provide one focal point for bringing MEPs together from different countries and parties. It should be recalled, however, that the European Parliament, despite the ongoing attempts to strengthen its hand in the policy process, remains poorly equipped to deliver significant policy reform in the environmental area.

The hardest task for ENGOs, however, is influencing the Council of Ministers. Voting patterns often follow the 'push/pull' formulae (Sbragia 1996) where Northern member states vote in favour of environmental legislation and poorer Southern members vote against it. The best opportunity for ENGOs at this stage is to coordinate lobbying across national members.

As Biliouri notes, 'The national offices play the most important role . . . [they] can approach national politicians, and through them, obtain allies within the Council of Ministers' (1999, p. 178).

However, once the policy has got this far it is too late to modify proposals before the Council, and the politics at this stage essentially centre on inter-state bargaining. Lobbying the Council directly is difficult because Council working papers are officially confidential – even though they are leaked extensively. Lobby groups often get most of their information from member states willing to share information with interest groups. In this regard, Council reports which contain redrafts of Commission proposals and cite the position of national delegations and their reservations on particular clauses are an essential lobbying tool in the hands of NGOs. These help to expose those governments who proclaim publicly that they are pushing for stronger environmental action, whilst stalling action behind closed doors. One other channel of influence is for NGOs to organise a press conference in the run up to a session of the Council as a way of expressing views about a draft directive and attempting to keep public attention focused on its progress in the hope that this will encourage the adoption of more far-reaching action.

Forms of influence

Besides general awareness-raising and pressure mobilisation among the public at the national level, NGOs can create policy problems by projecting ideas into the institutional arenas of the EU. It is at the stage of the policy process when a problem is being defined, expertise sought and the need for action discussed, that policy positions are developed. ENGOs therefore attempt to shape expectations about the nature of the policy that should be developed. Those NGOs that have the greatest role in shaping EU environmental law are those with significant reserves of human and financial resources, advance intelligence of policy proposals, good contacts and an ability to provide policy-makers with sound information and advice. Mazey and Richardson (1992, p. 110) note that 'Reputations for expertise, reliability and trust are key resources in lobbying in Brussels as elsewhere'. Euro-groups that are at once 'representative' and expert are those that are likely to be called upon most by the Commission. The expertise of environmental 'think-tanks' and policy research institutes (such as the Institute for European Environmental Policy) allows them to make an important contribution to the policy agenda at EU level (Grant, 1993a).

Richardson (1994) finds similarly in relation to EU water policy that ENGOs, strength is that 'They can out-match the big chemical companies in some respects simply because the groups access a different range of scientific expertise than even a large chemical company can command' (1994, p. 161). On the basis of this 'many of the other key players admit that they themselves are usually reacting to the agenda set by environmentalists' (ibid.). Environmentalists can create a 'megaphone' effect for scientific findings (Richardson, 1994), diffusing expertise to popular and policy audiences. It is less likely, however, that ENGOs will have the sort of expertise that the Commission demands at latter stages of policy formation. The levels at which emissions reductions will be set, the time-frames that will be employed and the extent to which technologies are in place that can realistically bring about the desired reductions are issues which industry groups, with more 'hands-on' experience, are more likely to be able to address than 'green' groups. The overriding concern of the Commission as policy develops, as the carbon tax debate bears out, will be the impact of the proposal upon the competitiveness of European industry rather than how adequately the policy reflects the severity of the issue implied by the science or the extent to which it proportionately responds to the degree of public concern surrounding the issue.

However, ENGOs do play an important role at the implementation stage of the policy cycle. 'Whistle-blowing' activity by environmental groups helps to notify the Commission of instances of non-compliance at the national level. Because these groups perform this function for the Commission, the Commission, in turn, is keen for NGOs to play an unofficial monitoring function in helping to bridge the 'implementation gap' between European-level legislation and national and local implementation. Chapter 6 illustrates this in practice. As part of this process, groups have also taken the opportunity to expose member states' inability to fulfil their commitments. Greenpeace International's report, 'The EU's next global warming factories' published in April 1994, showed how EU proposals to build new power plants would overwhelm the Union's goal of returning CO_2 to 1990 levels by the year 2000 (*Acid News*, 1994). Recognition of the importance of this NGO contribution is borne out by the fact that CNE was approached by the Earth Council, to produce a report assessing the effectiveness of EU climate policy in order to evaluate the EU's success in meeting the goals of the Convention (Newell, 1996a). The shift to voluntary agreements by industry, as part of the move towards flexible deregulation at the European level, will also create an increasing role for NGOs as watchdogs of corporate compliance with such codes. Instances of implementation failure will be reported to the Commission.

In considering the influence of NGO lobbying in Brussels, it helpful to discriminate between 'insiders' and 'outsiders'. Some NGOs are consulted or conferred a greater degree of access to the Commission than others. There are few settled and institutionalised patterns of consultation and the Commission is still in the process of developing its procedures of consultation and coordination, driven by the need to 'rationalise' the consultation process. In general, however, it seems that the following factors, taken together, help to account for the degree of influence that ENGOs have at the European level. One key determinant is the closeness of an NGO's relationship to principal decision-making bodies and actors at the European and national level. Climate Action Network-UK members, for example, have frequent meetings with UK representatives from Brussels and with the Secretary of State for the Environment, before meetings of the

EU Council of Ministers (Weir, quoted in Newell, 1997a).

Access is also a function of a group's lobbying style. The style of lobbying in Brussels serves to exclude NGOs that pursue a more confrontational campaigning style. Relations are preferred with groups willing to cooperate with the Commission without resorting to publicity to benefit their cause. There is still a perception among some Commission officials that environmentalists are obstructionist, anti-growth and overly reliant upon the media to attack decision-makers and companies (Mazey and Richardson, 1992, p. 126). To the extent that the Commission is able to set the terms of engagement with NGOs, some NGOs may be faced with a strategic dilemma whereby in order to gain 'insider' status with the Commission they have to drop their more confrontational approach to campaigning, whereas by doing this they may alienate their traditional membership base which is supportive of those tactics. This sort of dilemma is more likely to face groups like Greenpeace which attempts to combine insider access with 'outsider' strategies. Media-oriented public campaigns against the EU 'seed list', which Greenpeace was active in orchestrating, brought unwanted publicity to a controversial issue area, and are unlikely to endear the group to policy-makers who prefer more conciliatory participation. As writers in other issue areas have noted, however (see, for instance, Audley, 1997), where insider and outsider NGO strategies co-exist, policy-makers often make concessions to more conservative groups in order to reward positive engagement and, at the same time, to deter 'deviant' NGO behaviour. The irony, of course, is that 'insiders' would not make so much ground if it were not for the more threatening tactics of the 'outsider' groups. The fringe activity of groups like Greenpeace may benefit therefore those NGOs working within the EU system to affect change.

Because of this approach, however, Rucht (1993) argues that the Greenpeace office established in Brussels in 1988 is less effective at influencing EU decision-making processes than other groups operating at this level. The basis of his argument is that the organisation tends to stand apart from other alliances and is therefore less coordinated with other environmental groups in Brussels and 'less adapted to the task of lobbying and negotiating with EU bureaucrats' (1993, pp. 85–6). By contrast, groups such as WWF are said to have a close relationship with DGXI (Sbragia, 1996, p. 245). Hence, while WWF has set a limit of between 10 and 15 per cent on funding from public agencies, Greenpeace has abstained from asking the EU for money on the basis that this would compromise the group's autonomy of operation (Mazey and Richardson, 1992, p. 122).

When thinking about the political influence of smaller umbrella groups like CNE, it is important to remember that they are run by political entrepreneurs who spend much of their time fighting for the financial survival of their organisation (putting together grants, justifying their value to members and funder bodies) (Newell and Grant, 2000). Smaller groups constantly face difficulties financing the running costs of their offices which makes longer-term planning very difficult. To assume that such groups have the luxury to be able to reflect on a regular basis upon how best to exert influence, or how the changing architecture of EU politics may impact upon the way they operate as an organisation, is to misunderstand the day-to-day operation of small single-issue Brussels-based umbrella organisations.

It is not useful, however, to assess the influence and impact of ENGOs against the yardsticks we employ to gauge corporate influence where financial resources are paramount. Environmental groups play on policy-makers' perceptions that they represent [a] 'public interest' or at least that their motives transcend the narrow profit-making goals of their corporate counterparts. It is their perceived legitimacy in this regard, as well as the symbolic potency of their ideas and popularity of their values, that encourage policy-makers to take them seriously. The forms of leverage they exercise are altogether different, therefore, from those which business groups are able to operationalise.

Their influence operates in more subtle ways. It is about education, nurturing contacts and lobbying industry groups that may have more influence than themselves. The sorts of influence NGOs can lay claim to operate over the longer term, are less visible and are channelled in disparate ways so that assessing the overall effect of these various activities is a difficult task.

In general it is during the earlier stages of policy formation and agenda-setting and at the end of the policy cycle when proposals come to be implemented that ENGOs seem to have most impact upon the course and shape of EU environmental policy. This echoes Grant's (1993a, p. 44) conclusion that 'The most successful route for environmental groups to exert influence may be at either end of the policy process: influencing the construction of the policy agenda and highlighting implementation deficiencies.' Hence despite the fact that the Commission has stood up to powerful member states on environmental issues and provided the environmental lobby (or certain parts of it) with financial support, that the ECJ has come to be regarded as pro-environment in its interpretation of the law, and that the EP has displayed a strong environmental consciousness, environmental groups do not seem to have enjoyed a substantial and consistent degree of success in advancing their goals at the European level. This is inspite of Marks and McAdam's (1996, p. 114) claim that 'environmentalists have confronted a Union that, with the exception of the Council, has shown itself to be both attitudinally sympathetic and structurally open to the interests of the movement'. For these authors, the political opportunity structure offered by the EU is one that environmental groups should easily be able to adapt to. They note:

> there exists a real affinity between the tactics practised historically by the movement and the institutional openings afforded environmentalists by the emerging EU structure. Most environmental groups in Europe have been dominated by a combination of legal, electoral and lobbying strategies; the precise mix encouraged by the relative openness of the European Court, Parliament and the emerging policy community in Brussels. (1996:115)

And yet there is clearly a difference between being permitted formal access to the institutions in various formal and legalistic guises and being afforded the real decision-making authority to make a significant political impact that changes the EU's agenda. It also needs to be recalled that the Brussels policy style does not suit a number of environmental groups and is appealing only to those with the resources to fruitfully exploit the opportunities it affords.

One (ongoing) source of leverage for ENGOs is that the EU does seem to be keen to retain legitimacy as a leader on environmental issues as a way of raising its popularity with the European public and a channel for projecting leadership in international affairs. Continuing concern about the environment expressed in polls of public opinion (such as Eurobarometer surveys), together with the fact that these problems simply will not go away (and if anything will continue to get worse), mean that exogenous factors will also continue to ensure that the environment remains on the EU's policy agenda. The final factor which will ensure an ongoing role for ENGOs in the EU is the comparative weakness of the environment Directorate in the overall structure of environmental policy-making within the Union which means that it is constantly looking out for new allies to bolster its position and create constituencies of support for its policy agenda. This suggests it will continue to reach out to ENGOs, exploiting their expertise and access to popular constituencies.

There is no winning formula for success at Brussels level per se. As Rucht notes:

> Coherent, well-staffed and truly international organisations such as the WWF and Greenpeace have been quite successful according to their own criteria. But in relative terms, more informal, under-staffed and loosely coordinated transnational groups such as Climate Action Network, have also been successful. (1993, p. 91)

Conclusion

It has not been the intention of this chapter to argue that business groups always win and environmental groups always lose in the EU policy-making process. The picture is far more complicated than that. If it had not been for the presence and activity of environmental groups, EU environmental policy would not have developed to the extent that it has done. Nevertheless, business has won many of the key battles, notably over the carbon tax, an outcome which had an influence on the overall tone, content and direction of the debate about environmental policy.

Such outcomes are comprehensible if we reflect that business enjoys advantages along four key (interrelating) dimensions: (i) structural power; (ii) financial and staff resources; (iii) contact resources; (iv) legitimacy. In terms of the first dimension, Newell and Paterson observe (1998, p. 695):

> We have shown that the interests of fossil fuel companies have been consistent with the interests of 'capital-in-general' for much of the history of industrial capitalism, because growth in energy use is a precondition for general accumulation. As such, this creates a general background set of understandings where state managers assume that restricting energy growth restricts accumulation.

This ties in with the broader point that states, particularly in a context of global capital mobility, are sensitive to the fact that that they are structurally dependent upon those who possess capital (businesses) to achieve the overriding goal of economic growth (ibid.). Awareness of the nature of this relationship on the part of member states sharply constrains the menu of policy options particularly with regard to the environment where capital has key interests to defend.

We have noted in this chapter that the financial and staff resources available to business lobbies are much more substantial than those available to their environmental counterparts. In terms of contact resources, environmental groups remain heavily dependent on their relationship with DG XI, while business groups have contacts right across Commission services and at college of commissioners level, including the Commission President. Above all, because business cooperation and policy endorsement remains of vital importance to the economic growth objectives which lie at the heart of the European project, business perspectives carry more weight, are more useful in helping policy-makers to achieve their goals and are regarded therefore as more legitimate. Environmentalists can win policy advances, but they are often rather fragile and subject to dilution at the implementation and enforcement stages. The asymmetry of power between business and environmental groups should not be a surprise in an organisation whose central purpose has been to create and develop an internal market.

3
Implementation and Enforcement

Introduction

European Union (EU) environmental policy is intended to have particular outcomes. Achieving these outcomes requires EU environmental law to be implemented and enforced effectively in the member states. This is the final stage of the 'loop', which involves policy being conceived, drafted in legislative form, adopted as EU legislation, and implemented and enforced by the member states. It is only by observing the entire loop that the impact of policy decisions can be fully evaluated. No matter how appropriate the design and form of a legislative instrument may be, objectives will not be met unless EU environmental legislation is correctly implemented into national law, then applied and enforced effectively by the competent national authorities. This, in turn, can have profound implications for the design and content of EU environmental policy in the future.

The legislative dimension of EU environmental policy has developed rapidly since the First Action Programme on the Environment in 1973. There are now over 200 separate legislative measures designed to ensure a high level of protection for the environment (European Commission, 1996c, p. 1b). Yet despite twenty-five years of coordinated policy activity to protect the environment, and some evidence that the existence of EU environmental policy has assisted the 'ratcheting up' (Golub, 1996) of environmental standards in some member states where earlier environmental measures were ineffective or non-existent, the overall impact of

EU environmental policy through the legislative route
been a notable success.

In 1993, the Fifth Action Programme on the Environment ide
tified a slow but relentless deterioration of the general state of the environment, notwithstanding the measures taken as regards climate change, acidification and air pollution, depletion of natural resources and biodiversity, depletion and pollution of water resources, deterioration of the urban environment, deterioration of coastal zones and waste (European Community, 1993, p. 5). Furthermore, by the Commission's own admission, there are serious weaknesses in the current state of implementation and enforcement of environmental law in most parts of the Union (European Commission, 1996c, p. 2) which mean that many of the environmental policy decisions taken at EU level have not had the intended impact.

The Fifth Action Programme on the Environment (European Community, 1993, p. 80) identifies a number of reasons for these deficiencies in the implementation and enforcement of EU environmental measures. First, there has been a lack of overall policy coherence, partly due to an evolving, sometimes shifting, agenda as the scope of environmental policy grew, and partly because much of the environmental legislation grew in an ad hoc manner. Second, there has been a narrow choice of policy instruments, which has tended to rely on regulation through 'command and control' measures (Scott, 1998, p. 24) rather than the more 'advisory' approach that is discussed later in this chapter. Third, in the past there has been a need for unanimous agreement in the Council of Ministers before environmental legislation could be adopted. This often resulted in political compromises that were difficult to translate into practical environmental measures. Fourth, the type of legal instruments used (Directives) have often given rise to difficulties in their incorporation into national law, while the complexity of the subject matter has often given rise to problems of interpreting exactly what is entailed by implementation of EU environmental policy objectives through legal obligations placed on member states. Fifth, differences in the practical application of environmental legislation by national, regional and local competent authorities has led to problems of enforcement.

This chapter examines each of these reasons for deficiencies in cement in more detail. It also addresses e EU can achieve effective implementa- forcement of environmental law capable vironmental policy has its intended im- tions will help to clarify the nature of the

ne overall approach of the book, for the pter we will use the terms 'European Union' and ... e for the more legally correct terms 'European Community' an... EC' in the study of environmental law. This terminology is used for the sake of consistency and simplicity within the volume as a whole, but we do recognise the finer points of the nuances between the EC and EU in this respect. In the context of setting out definitions in this chapter, implementation means legislative, regulatory or binding administrative measures taken by a national competent authority to incorporate the obligations, rights and duties enshrined in EU environmental law into the national legal order. Practical application means the incorporation of EU law into individual practical decisions by the competent national authorities – for example, by putting in place a plan to implement environmental improvement measures. Enforcement means steps taken by competent national authorities to ensure compliance with legislation. These steps might include, for instance, environmental monitoring, fines for breaching environmental standards and compulsory corrective measures to be taken. While these definitions differ from those used by some other commentators on environmental law, they do accord with those used by the European Commission (1996c, p. 1b) and provide a useful starting point for discussing the problems currently facing EU environmental policy.

Characteristics of EU environmental legislation

The implementation and enforcement of legislation remains the weakest link in EU environmental policy (Somsen, 1996, p. 198) for a number of reasons that relate to the particular character of environmental problems.

Environmental protection has to take account of complex interrelationships between air, soil and water and biodiversity. The

Commission is well aware that, unless care is taken, action to protect one medium can adversely affect another (European Commission, 1996c, p. 3) and the Treaty acknowledges the complexity of different environmental problems in Article 174(2) (formerly Article 130r(2) of the Treaty) which states that 'policy on the environment shall aim at a high level of protection taking into account the diversity of situations in the various regions'.

Once environmental degradation has occurred, its effects often take years to rectify and are in some instances irreversible. This means that the effective implementation and enforcement of environmental law is often a more immediate priority than for some other EU policy areas where it may be easier to rectify the damage caused. Moreover, Krämer (1997, p. 1) has suggested that the reason why EU environmental legislation is often inadequately implemented and enforced is because it seeks to achieve an aim – the protection of the 'environment' – which is so general that it often lacks the commitment of specific policy communities to definable policy outcomes in the same way as are found in other EU policy areas such as competition policy or internal market policy. The environment therefore comprises a set of issues that do not engender what the Commission would call a 'proprietary stake' (European Commission, 1996c, p. 3) with benefits that will directly accrue to specific policy actors. The benefits of clean air, soil and water and biodiversity do not accrue solely to particular groups in society.

In many ways, the implementation and enforcement problem for EU environmental law can thus be explained by crude cost–benefit analysis: although the environment as a whole would benefit from reduced levels of pollution and higher environmental standards, there remain problems about who should be responsible for those improvements and who should bear the costs. Specific problems arise because the most frequent polluters – in business and agriculture – have few direct incentives to comply. They perceive environmental legislation as invariably imposing costs in terms of cleaner production methods, lower productivity or repairing damage caused to the environment. There is also the common complaint that there is no 'level playing field' for the implementation and enforcement of environmental law throughout the EU, with the result that competitors in other member states are not required to bear the same costs of compliance (Dehousse et al., 1992).

So, despite a widespread consensus that effective implementation and enforcement must be ensured, no agreement has yet been reached on how this objective should be met. Consequently, there remains an 'implementation gap' (Rehbinder and Stewart, 1985), or an 'implementation deficit' (Jordan, 1999a, p. 71), and an 'enforcement gap' (Somsen, 1996, p. 198) in environmental law that undermines the intended impact of EU policy in this area.

This chapter explores what is entailed by implementation and enforcement in the context of environmental policy and considers how improvements can be made to the principal mechanisms currently used to ensure that the EU's environmental policy goals are achieved.

Implementation

Implementation, the process of giving effect to EU environmental policy at national level, is achieved when EU legislation is enacted into national statute and administrative practice. The obligation to implement EU law is found in Article 10 (formerly Article 5) of the Treaty, which requires member states to 'take all appropriate measures, whether general or particular, to ensure fulfilment of the obligations arising out of this Treaty or resulting from action taken by [the] institutions'. In practice, this obligation has three components: the establishment of rights and obligations as laid down in the text of legislation; the amendment of contradictory national legislation; and the creation of the necessary structures to ensure that the terms of legislation are carried out.

In assessing the effectiveness of national measures to ensure the implementation of environmental policy measures in the EU, particular attention needs to be focused on the type of legal instrument used. Although Directives are the most common means of enacting environmental policy, the Council and Commission are also both empowered to issue Regulations which, once adopted, automatically become part of the national legal framework in each member state without the necessity for legislative or administrative implementation. Yet while, on the face of it,

Regulations are directly applicable they do often require member states to take action to ensure compliance with their provisions. A Directive adopted by the Council of Ministers, on the other hand, requires member states to implement national laws, regulations and administrative provisions necessary to bring national legislation into line with EU law.

Once a Directive has been implemented into national law, it is assumed that the practical means by which its aims will be achieved are not only via the Directive, but also through the national implementing legislation. The intention is that a Directive will allow for national diversity and variation within the permitted scope of the Directive's text. It is not expected that every member state will have identical implementing legislation to achieve the same aims. Directives impose obligations of the result to be achieved by a choice of means, while Regulations impose obligations of form, in the sense that a Regulation automatically becomes an integral part of the national legal system (Freestone and Davidson, 1988).

In allowing member states to choose the exact form in which it is transposed into national law, Directives therefore offer national governments considerable discretion and since over 140 EU environmental measures are in the form of Directives, this also raises a number of important national implementation issues. To implement a Directive correctly, a member state must ensure that national legislation complies with it fully and within the designated time limit. Once a Directive has been adopted as EU law, member states are then normally given between 18 and 36 months to implement (or transpose) it into national law. Each member state is then normally required to send a 'compliance letter' to the Commission, with notification of the measures that have been taken to implement the Directive. The Directive itself will often specify what information the compliance letter is required to contain. However, the track record of member states correctly notifying the Commission of measures to implement Directives has not been good.

The legal, technical and administrative measures used by member states to implement environmental Directives into national law vary considerably, according to legal traditions and existing national procedures. The resulting problems of ensuring uniform effectiveness of environmental law in the EU are well known. A member state may implement a Directive into national law late, not at all, may do so only partially or inadequately. In many instances, implementation of EU environmental legislation into national law has been slow. Late implementation of Directives into national law is the norm, not the exception and it is rare that more than three member states have notified their implementation measures to the Commission within the specified time limit (Krämer, 1997, p. 7). By the end of 1997 member states had notified implementing measures for nearly 97 per cent of environmental Directives (an improvement on the comparable figure of 91 per cent in 1995), but in some member states this left as many as 18 Directives still awaiting implementation (European Commission, 1998e, p. 95).

Furthermore, although more recent figures are not available, in 1995 the Commission registered 265 suspected breaches of EU environmental law, based on complaints from the public, Parliamentary questions or petitions and cases detected on the Commission's own initiative. This accounted for over 20 per cent of all infringements of EU law registered by the Commission that year and, by October 1996, over 600 environmental complaints and infringement cases were outstanding against member states, with 85 awaiting rulings by the Court of Justice (European Commission, 1996c, p. 2). However, these figures may actually only be the tip of the iceberg. Since the Commission is almost entirely reliant on cases of non-implementation with environmental law being brought to its attention by concerned parties, the actual size of the implementation problem is undoubtedly much larger and has been described by Weiler (1991) as a 'black hole' in EU law, representing an obstacle to the credibility of the whole Union (Anderson, 1988).

The concern that Directives are not properly implemented was acknowledged by the member states themselves at the Maastricht Intergovernmental Conference, where they issued a Declaration which stated that they recognised that it is central to the coherence and unity of the process of European construction that each member state should fully and accurately transpose Directives into national law addressed to it within the deadlines laid down. In addition, while recognising that it must be for each member state to determine how the provision of EU law can best be enforced in the light of its own particular institutions, legal system and other circumstances, the member states considered it essential for the proper functioning of the Union that the measures taken by the different member states should result in EU law being applied with the same effectiveness and rigour as in the application of their national law (European Community, 1992b).

The Maastricht Declaration therefore enjoined member states to transpose Directives fully and adequately into national law within the specified deadlines and recognised that, while member states might take different measures to enforce EU law, these should result in measures being applied with the same effectiveness and rigour as national law. This latter point raised the related issue of enforcement.

Enforcement

Assuming that EU environmental law is implemented into national law in an appropriate manner, the problem then becomes one of enforcement for the competent national, regional and local authorities that are responsible for monitoring the practical application of environmental standards in the member states.

The enforcement mechanisms used will depend on the administrative structures in place in the member state concerned. Where member states provide for enforcement to be achieved at regional or even local level, geographical variations may result (Krämer, 1997, p. 8). In Germany, for example, enforcing EU environmental law, and introducing appropriate implementing regulations, is often the responsibility of the Länder. EU law may then be applied and enforced differently and, on occasion, incorrectly in different parts of Germany. Similarly, in the United Kingdom, responsibility for enforcement is divided between the competent authorities in England and Wales, Scotland and Northern Ireland. Decentralisation of the enforcement process adds to the complexity of ensuring the effectiveness of EU environmental law since it increases the possibility of incomplete or incorrect implementation and enforcement, even within different regions of the same member state. This raises important issues because, in a decentralised system, the national government is not necessarily the body responsible for enforcement and, according to Bell (1997, p. 67), may not be in a strong position to ensure high standards of enforcement in all parts of that member state.

So, it is clear that national administrative frameworks for enforcing EU law differ, sometimes in important respects. The role of EU institutions in ensuring the effectiveness of national enforcement systems relative to one another therefore becomes extremely difficult (see Sutherland, 1992; Dehousse et al., 1992; Krislov et al., 1986).

Role of EU institutions in ensuring implementation and enforcement

Variations in the implementation and enforcement of EU law are monitored by the Commission, which may then bring actions for failure to implement and enforce legislation before the Court of Justice (see Scott, 1998, pp. 148–55). The Commission undertakes this task in accordance with its role as guardian of the Treaty. This obligation is set down in Article 211 (formerly Article 155) of the Treaty and relates not only to ensuring that EU law is correctly implemented into national law, but that it correctly 'applies' in specific situations (Krämer, 1997, p. 5).

To do this, the Commission uses three criteria: whether a member state has implemented the EU provision into national law and has issued the Commission with a compliance letter; whether implementation has been achieved for the whole of the member state's territory in a manner which is correct and complete; and whether there are national enforcement provisions to ensure that an EU provision is actually applied in a specific situation (Krämer, 1991, p. 9).

Where steps taken by a member state to implement and enforce EU environmental law are considered inadequate, the Commission may begin a two-stage infringement procedure: a preliminary administrative stage and a judicial stage. The preliminary administrative stage commences with the Commission writing to the member state informally, asking it to explain how the measures it has taken are sufficient to comply with EU law (Bell, 1997, p. 68). If a satisfactory answer is not received, a formal letter will be sent and a response from the member state will be required. If a satisfactory response is still not received, the Commission will issue a Reasoned Opinion, setting out why it considers that the steps taken by the member state are inadequate. By the end of the preliminary administrative stage, most cases have normally been resolved to the satisfaction of both parties.

Where the matter is not settled at the administrative stage, the Commission may decide to exercise its power to bring judicial infringement proceedings before the Court of Justice under Article 226 (formerly Article 169) of the Treaty. Another member state or a third party may join in Article 226 proceedings as third party to argue on behalf of either party. Bell (1997, p. 68) points out that the UK did this to support the Commission in the *Danish Bottles* case. If the Commission chooses not to bring infringement proceedings, another member state may do so under Article 227 (formerly Article 170) of the Treaty, although in practice this is very rare. Third-party individuals have no right to bring infringement proceedings before the Court of Justice. They may merely bring the alleged infringement to the attention of the Commission.

The Article 226 procedure itself is not transparent – that is to say there is no public discussion of the case. If an individual makes a complaint which results in a case being brought by the Commission, the complainant will normally be informed of the Commission's decision to commence proceedings under Article 226 or not to pursue the case. The complainant may also be given a brief summary of the reasons that led to that decision, but has no right to a full explanation of the Commission's reasoning and may well receive no information at all (Krämer, 1997, p. 7). This has led to concerns that the Commission may be influenced by political considerations, such as the character of the complainant, rather than simply deciding whether to prosecute a member state on the facts of the case.

Assuming an Article 226 case is pursued by the Commission, the Court of Justice then becomes the ultimate arbiter of whether a member state has correctly implemented and enforced EU environmental law. In the past, the Court had few powers to enforce its ruling (Bell, 1997, p. 68). This resulted in a worrying trend of member states failing to comply with a judgment of the Court of Justice once an Article 226 case had been concluded.

In an attempt to improve the effectiveness of the Article 226 procedure, the Maastricht Treaty on European Union introduced the possibility of fines being imposed against member states under Article 171 (since the Amsterdam Treaty renumbered as Article 228). Under what is now the Article 228 procedure, the Commission may issue a Reasoned Opinion if it considers that a member state has not complied with a ruling of the Court of Justice. If a member state then continues to ignore the ruling, the Commission can bring the case back before the Court of Justice, which may then impose a financial penalty on the member state. For a case involving failure to implement environmental measures, the size of the financial penalty will depend on the seriousness of the case and will take into account factors such as the severity of the environmental damage caused by the breach of environmental law (Bell, 1997, p. 68). The duration of the violation of EU law is also taken into account in setting the level of the fine, as is the cooperation of the member state in seeking to resolve the matter after the initial Court of Justice case.

The level of fines for failure to implement and enforce EU environmental law will also depend on the member state's voting influence over the legislation that has been breached and that state's gross domestic product. Bell (1997, p. 69) reports that fines imposed for a breach of environmental Directives have been well in excess of 500 000 ECU in some cases and that the Commission has also considered withholding EU funding for environmental programmes as an added deterrent.

Although the Commission has made considerable use of Article 228 to impose sanctions on member states, this appears only to have highlighted the size of the implementation problem rather than improving the situation. In recent years there has been a marked increase in the use of Article 228 against member states for failure to transpose environmental Directives into national law, despite an Article 226 judgment by the Court of Justice (Krämer, 1997, p. 8). Another drawback is that, while the Article 226 procedure allows a formal mechanism by which member states can be brought before the Court of Justice for infringements of EU environmental law, it is a lengthy procedure. It may take several years before a case is heard by the Court. Furthermore, Article 226 is not designed specifically with environmental law in mind. It is a means of reacting to decisions and actions after they have occurred and, even if the outcome is that EU law is better applied as a result, Article 226 is not a mechanism for preventing degradation or damage to the environment from occurring in the first place (Somsen, 1996, p. 198).

Article 226 is also of limited use as a mechanism for enforcing environmental law because Directives and Regulations have to be applied on a daily basis by a large number of competent authorities in the member states. It is an impossible task for the Commission to monitor whether these competent authorities are correctly implementing and enforcing the environmental obligations of EU membership. Nor does the Commission have specific inspection or control responsibilities for environmental issues. The Commission does not have the resources to make frequent visits to investigate suspected breaches of environmental law, while the European Environment Agency has no powers at all to investigate the practical application of environmental policy in any specific case.

Even if the Commission were able to detect a larger number of infringements, it would be impossible for the Court of Justice to then prosecute for infringements under Article 226 because of the sheer number of cases this would involve. Since Article 226 is designed to bring infringement actions against national governments, it would also be judicially impossible for the Commission to bring actions directly against the competent authorities implementing EU environmental law at a local level.

Using judgments of the Court of Justice and the national courts to ensure the effective implementation of EU environmental law also has several disadvantages which flow from the way that the system has evolved. Foremost amongst these is that, by definition, judicial rules have been developed in an ad hoc way, less coherent and less comprehensive than a legislative framework. Litigation rates are influenced by diverse national legal cultures and differ substantially among member states. This makes it unlikely that litigation alone would result in the uniform effectiveness of EU law (Snyder, 1990) and, in any case, as mentioned earlier, the Commission simply does not have the administrative systems available at the national, regional or local level which could inform it about the application of EU environmental standards (Krämer, 1997, p. 6).

The Commission recognises that different actors (the Commission, member states, regional and local authorities, industry, citizens and non-governmental organisations) each play an important role in the 'shared responsibility' of ensuring the effective implementation and enforcement of EU environmental law (European Commission, 1995). But this approach itself raises practical problems because of the wide disparity in the approaches to environmental inspection taken in different member states.

In cases of large, well-publicised environmental accidents, such as pollution of the Rhine, breaches of EU law may well be self-evident. However, in monitoring the day-to-day implementation and enforcement of environmental law, the Commission is almost entirely dependent on information supplied to it on an ad hoc basis through complaints by citizens or non-governmental organisations (NGOs), petitions to the European Parliament, oral questions from the European Parliament, reports from the media and information supplied by member states themselves. These complaints constitute the main route by which the Commission obtains knowledge of presumed or actual non-application of EU environmental law, although it appears that the number of complaints has decreased slightly while Parliamentary questions have increased (Krämer, 1997, p. 6). Petitions submitted to the European Parliament are also forwarded to the Commission for appropriate follow-up rather than being dealt with by the Parliament itself.

Although this information is all extremely useful to the Commission, it is provided in a piecemeal fashion and is largely unverifiable. The detrimental consequences of poor monitoring procedures and lack of reliable data on the EU's environment have been acknowledged by the Commission (European Commission, 1996c, p. 5). Without reliable and comparable data being available from the member states, it is impossible to assess environmental problems properly. This problem was addressed with the establishment of a European Environment Agency (EEA) which, since 1993, has acted as a EU-level institution to collect, collate and report on environmental information provided by the member states so that data can be presented in a form that will allow environmental conditions in the member states to be more easily compared. This approach consolidates earlier steps towards the harmonisation of reporting requirements in the 1991 Environmental Reporting Directive (91/692/EEC).

Krämer (1997, p. 7) has described how the Commission goes about investigating environmental complaints. The Commission retains discretion over which complaints and petitions it wishes to investigate and, in practice, does not have the time or resources to investigate all cases brought to its attention. Where a complaint is investigated, the normal procedure is for the Commission to confront the member state with the information submitted and ask for a response and, where appropriate, supporting documentation from the national authorities. The Commission rarely has the resources to investigate the case thoroughly on its own account.

One innovative approach that the Commission has taken to investigate complaints are the so-called 'package meetings' (Krämer, 1997, p. 7), whereby a number of complaints and suspected breaches of EU environmental law are discussed with the appropriate national authorities. The Commission will examine the facts put before it and, without seeking to enter into litigation with the member state concerned, look for a solution that will achieve the necessary environmental improvements. Krämer reports that Italy, Spain and France even admit to these meetings representatives of regional and local authorities and allow them to enter into discussions with the Commission. This procedure is not followed by other member states, who restrict the dialogue to one between central government and the Commission.

Role of member states in ensuring implementation and enforcement

Under Article 249 (formerly Article 189) of the Treaty, in conjunction with the general principles set out in Article 10 (formerly Article 5) of the Treaty, member states are under a duty to take all necessary legal and administrative steps to ensure the fulfilment of the obligations arising out of EU legislation, while Article 175(4) (formerly Article 130s(4)) of the Treaty specifically requires that member states 'finance and implement the environment policy'. It should also be remembered that, under Article 176 (formerly Article 130t) of the Treaty, member states are allowed to introduce or maintain tighter standards than are set down in EU environmental law.

The obligation to introduce standards that are at least as stringent as EU environmental law applies equally to measures necessary to implement Regulations and Directives. Although Regulations are theoretically directly applicable in the member states and should not require implementation into national law, in reality the EU does not have the resources to enforce regulations itself so there is little practical difference between the obligation of both types of measures, except that Directives require formal implementation via transposition into national law.

Even following implementation into national law, the interpretation of EU legislation depends on national courts if there is uncertainty as to the precise legal rights and obligations. Article 234 (formerly Article 177) of the Treaty provides a mechanism for national courts to refer problems of interpretation to the Court of Justice via the preliminary ruling procedure. Under Article 234, the onus is on national courts to seek clarification on a point of law from the Court of Justice and, since the role of the Court of Justice under the Article 234 procedure is restricted to one of interpretation, not application of EU law, the effectiveness of environmental law is dependent on the national courts referring the appropriate cases where clarification on implementation is required.

Member states must not only implement and interpret the provisions of Directives into national law; they must also take the necessary steps to ensure the practical application of the law – ensuring, for example, that drinking water supplies or bathing waters comply with the standards set by Directives. Failure of the member states to ensure the practical application of EU environmental legislation into national law has long been regarded as a major hindrance to an effective environmental policy (see, for example, Collins and Earnshaw, 1992; Crockett and Schultz, 1991; Krämer, 1991; Siedentopf and Ziller, 1989). Even the member states themselves recognise that this deficit in application of environmental law (Krämer, 1997, p. 2) is largely due to their own failure to adequately provide for the practical application of EU legislation.

Vogel (1986) has attributed differences in practical application and enforcement to variations in styles of enforcement in the national setting. On the basis of Vogel's analysis, one would expect the way in which EU environmental law is enforced in each member state to vary from one another for reasons relating to the organisation of enforcement bodies and the style of enforcement activity adopted in that national setting. Two principal styles of environmental enforcement can be identified: legal action (the coercive approach) and advice and persuasion (the advisory approach).

National and regional environment inspectorates have a number of statutory powers which may be used to sanction businesses or individuals for behaviour that does not comply with environmental law. These sanctions may include referring cases for prosecution, issuing fines or closing a commercial premises. The coercive model is therefore characterised by the rigorous application of rules, together with a formal enforcement style that follows the letter of the law. Relying on legal procedures akin to a coercive model to achieve compliance has been described by Hawkins (1984) as a 'sanctioning strategy'. Hutter (1989) even suggests that the number of prosecutions may in some instances be regarded by an inspectorate as the measure of policy success in themselves, regardless of their relationship to improved environmental standards.

The advisory or persuasive approach, on the other hand, emphasises the provision of information on how compliance with regulatory requirements can be achieved, with legal sanctions used only where there is a high risk of pollution accidents, where pollution has already occurred or where the provision of advice does not achieve the desired outcome of ensuring compliance with environmental legislation. The advisory approach to environmental enforcement can be subdivided further by making the distinction between the persuasive approach and the insistent approach. Hutter (1989) characterises the persuasive approach as a situation where an inspector explains the legal requirements and discusses with an employer how improvements can be best achieved. An employer is then given the opportunity to rectify problems without the threat of prosecution being imminent. The insistent approach, on the other hand, relies on an environmental agency setting out the improvements to be made by a polluter within a clearly defined time-scale, insisting on the ends to be achieved more often than insisting on the means of achieving those outcomes. Where improvements are not made, the coercive approach will be brought into play to ensure compliance.

The enforcement style most often favoured by environmental agencies in the member states is the advisory approach (see Richardson, Ogus and Burrows, 1983), characterised by a cooperative, conciliatory approach by enforcement officers who seek to achieve compliance through persuasion, negotiation and public education or awareness campaigns. The advisory approach should not, however, be seen as an alternative to the coercive approach. Where companies seek not to comply with the law, either because of the high cost of complying with environmental legislation, or because they do not anticipate that the threat of prosecution is real, the coercive approach will be available to ensure even-handed and effective enforcement of EU environmental law. Above all, the threat of sanctions must be real and be used where appropriate.

Co-ordination of implementation and enforcement

The Commission has repeatedly stated its intention to systematically monitor the effectiveness with which national enforcement agencies apply EU environmental standards, particularly through the European Environment Agency (EEA). The EEA was established under Council Regulation (EEC) 1210/90 of 7 May 1990. The Regulation came into force when discussions on where the Agency should be based were concluded in October 1993.

The EEA is now based in Copenhagen and has been set up with the specific task of gathering and disseminating information on the state of the environment in the EU. In order to assist the EEA in this task, the European Environment Information and Observation Network were established, with member states required to inform the EEA of their main environmental information networks. The EEA is then expected to incorporate national environmental networks into a single EU network (Bell, 1997, p. 63).

The EEA does not have enforcement powers, despite attempts by the European Parliament to include such powers in the

Regulation which established the Agency. However, the debate that took place over EEA enforcement powers is reflected in a provision in the Regulation that requires the Council to reconsider the scope of the Agency's powers – in particular, specifying a possible role in monitoring the implementation of EU environmental legislation (Bell, 1997, p. 63). The Agency's role may, therefore, yet evolve into an inspectorate capable of enforcing EU law in the member states.

Steps have also been taken to encourage cooperation between national environmental enforcement agencies. In 1992 the 'Chester Network' was formed to informally exchange information and experience on environmental standard-setting and enforcement (European Commission, 1996c, p. 18). Then, in 1993 the Fifth Action Programme on the Environment (European Communities, 1993) proposed the setting-up of an implementation network comprising representatives of the Commission and the environmental enforcement agencies of the member states. The Chester Network was subsequently modified into the Informal Network for the Implementation and Enforcement of Environmental Law (IMPEL). The role of IMPEL is to consider the implementation of EU environmental legislation, particularly relating to measures that can be taken to improve its enforcement by national, regional and local authorities (Bell, 1997, p. 63).

However, Dehousse et al. (1992) have expressed doubts that agencies and networks such as the EEA and IMPEL are actually well equipped to address the implementation and enforcement problem. They suggest that what may actually be required are EU-level agencies with substantive enforcement powers and that, although some pressure may be imposed on member states to more rigorously enforce law through EU-level monitoring and issuing reports on national enforcement regimes, this may not be enough to encourage national enforcement agencies to take a firmer line. But the suggestion that some form of European environmental protection agency should be created has not been enthusiastically received, particularly since it would involve issues of national sovereignty if national enforcement powers were ceded to an EU institution. At present, EU-level agencies, such as the EEA, only play a role in the collection and dissemination of information throughout the Union.

The increasing penetration of EU-level institutions into national legal and administrative frameworks has created a new institutional environment. Yet, ensuring the effective implementation and enforcement of EU environmental law still involves basic issues regarding the legitimacy of the European Union and the role of subsidiarity in apportioning responsibilities to the most appropriate administrative level. Consideration of how to improve the effectiveness of environmental policy leads to a wider debate over the role of the EU institutions and their future development. In this respect, the institutional roles of the Commission, in conducting administrative negotiation, and the Court of Justice, in developing a judicial liability system, are crucial. These tasks of administrative negotiation and judicial control are likely to be developed in tandem to ensure that, at the national level where implementation and enforcement of environmental law takes place, competent authorities are doing sufficient to ensure the effectiveness of EU environmental policy.

In ensuring the effectiveness of EU environmental policy, the current system has worked best where it has created legal principles and cooperated with national administrations. Through adjudication and negotiation, relations between the European Union and its member states have followed a path of incremental development. Yet neither the Commission nor the EEA has sufficient resources or powers to ensure the effectiveness of EU environmental policy without the cooperation and the commitment of competent authorities responsible for implementation and enforcement at the national level. Consequently, national environmental agencies remain crucial to ensuring the effectiveness of EU environmental policy goals.

Commission proposals to improve implementation and enforcement

In October 1996 the Commission published its own suggestions to improve the implementation and enforcement of environmental law (European Commission, 1996c). The Communication outlined the Commission's role in making recommendations to assist member states in carrying out environmental inspection tasks, particularly by drawing up guidelines, thereby reducing the currently existing wide disparity among member state inspections. The Commission also said that it would consider whether there is a role for an EU body, presumably the EEA, with powers to audit national environmental reports.

The Commission acknowledges the increasing number of environmental complaints, either made directly to the Commission by citizens or environmental NGOs, through petitions made to the European Parliament, or through written and oral questions made by MEPs. However, in the Commission's view, one reason for the increasing number of complaints is that there are misunderstandings about what environmental legislation actually requires member states to do and the public are often not aware that clarification of implementing measures is available directly from national authorities. The Commission is therefore considering setting up guidelines to help complainants understand what the proper procedural mechanisms are for handling environmental complaints, particularly the fact that, in the first instance, many complaints should be dealt with by local or national authorities in the member states.

Where complaints are not dealt with through administrative procedures, legal actions are available as a last resort for solving implementation disputes. The Commission is aware that access to justice for citizens or environmental NGOs through the national courts is not always available and that these complaints often have to be made directly to the Commission itself. It is therefore considering the need for guidelines, reminding national courts that citizens and representative organisations do have a role to play in encouraging the application and enforcement of EU environmental legislation at member state level. The Commission is also keen to take steps to ensure that all proposals for new environmental measures or amendments of existing legislation are clearly worded in order to make legislation as accessible as possible to citizens and to make the implementation process simpler and quicker. Furthermore, the Commission is aware of the need, when formulating environmental legislation, for a full and open consultation procedure with EU institutions, environmental experts and the public in member states. The consultation exercise that preceded Commission proposals for a framework Directive on water policy, for example, is described in Chapter 6 and the Commission intends to expand this type of pre-proposal consultation exercise for future environmental policy-making.

Consultation about precisely what form national implementing measures should take is also important to ensure the effectiveness of EU environmental policy. The Commission has acknowledged that proposals for environmental legislation are likely to affect a wide range of actors in the member states, who should be consulted before implementation takes place. Failure to consult simply means that those affected are either unaware of new legislation or feel that it does not reflect their needs and concerns. Because lack of awareness of environmental legislation can itself be an impediment to implementation and enforcement, the Commission is now keen that member states ensure that they have the appropriate mechanisms in place to enable proper consultation prior to the adoption of national transposing measures. To assist member states in their task, the Commission has suggested that national governments should make better use of the Explanatory Memorandum which accompanies EU legislative proposals, since this document outlines the scope of the proposed measure and assesses its likely impact in the member states. In addition, the Commission has said that it will consider giving financial and technical assistance to member states where this will increase awareness of environmental legislation, with a particular emphasis on improving the knowledge of judges, lawyers and government officials in interpreting obligations under EU law.

However, even when environmental measures are clearly worded and widely understood, there will still be those who choose to ignore their legal obligations to curb the environmental damage they are causing. The Commission has recognised that EU environmental measures must require member states to put in place appropriate sanctions so that there is a sufficient threat to deter non-compliance with environmental legislation. These sanctions may be in the form of administrative penalties, such as the withdrawal of pollution consent permits, civil penalties in the form of fines, or penal penalties in the form of custodial prison sentences. The Commission has also suggested that both member states and the EU institutions should publicise the types of sanctions that polluters will face in order that the deterrent effect of these measures is used to maximum effect.

Much criticism has been made of the lack of reliable data that would enable better EU-level monitoring of national implementation and enforcement of environmental legislation. The Commission has announced that its annual report on monitoring the application of EU law, which covers all aspects of EU legislation implemented in the member states, will be expanded to cover national measures to apply environmental law in much greater detail. Furthermore, the Commission is considering how implementation can be improved by using the informal IMPEL network to encourage cooperation between different national environmental enforcement agencies. It has said that it will make proposals for improving, developing and reorganising IMPEL's tasks, particularly by encouraging the creation of national coordination networks linked with IMPEL. The Commission has also said that it hopes the 1991 Environmental Reporting Directive will ensure that the best possible information is available on the effectiveness of EU environmental measures and can be used in the formulation of its policies on environmental protection. However, the first reports will take some time to become available and the Commission has announced that, in the meantime, in order to add to the available information it will launch case studies to assess the implementation, application and enforcement of selected provisions of EU environmental law in the member states. The European Environment Agency will clearly play an important role in collecting and interpreting information on the state of the environment in Europe and the Commission committed itself to closely involving the EEA formulating proposals for new EU environmental measures and reviewing existing legislation.

Finally, the Commission has announced that it will give greater consideration to how the EU can help fund environmental projects in the member states. In 1996, it gave financial assistance of 1.7 billion ECU to assist with environmental improvements and it will consider increasing this amount further, where it considers this appropriate (European Commission, 1996c, p. 23). The Commission has linked the prospect of additional financial assistance for the environment with an announcement that it will not fund projects, such as new road building in the member states, which are likely to have negative effects on environmental impacts which are specifically protected under EU law, such as those protected under the Habitats Directive, unless the project complies with EU environmental standards.

Conclusion

This chapter has shown that the current means of overcoming the implementation and enforcement problems associated with ensuring the effectiveness of EU environmental policy relies principally on actions brought by the Commission before the Court of Justice under Articles 226 and 228 of the Treaty. We have seen that this reliance on a judicial approach to ensuring effective implementation and enforcement has proved insufficient as a means of protecting the environment. Deficits in the implementation and enforcement of EU environmental legislation remain widespread and need to be addressed by new approaches and new policy instruments. The Commission's 1996 Communication on implementing environmental law in the EU has suggested how a new approach might be undertaken. The Commission's recommendations imply that the improved implementation and enforcement of EU environmental legislation is likely to rely on a broader mix of environmental instruments, including steps to improve public consultation during the policy formulation stage, providing better information and advice on how implementation and enforcement should be achieved, co-ordinating the activities of national environmental enforcement agencies, and collecting more reliable information on the state of the environment. The Commission has recognised that new and innovative approaches to implementing and enforcing legislation will be needed if EU policy is to have its intended impact on Europe's environment. In doing so, the Commission has acknowledged that the absence of effective implementation and enforcement remains a major flaw in the EU environmental policy process.

4

The EU as an External Actor

The relations *between* member states of the EU over questions of environmental policy form just one part of a complex maze of policy interactions that help to constitute the foreign environmental policy of the EU. This chapter will be mainly concerned with the EU's relations with other states in the global arena in relation to the environment. But in order to make sense of these relationships, continual reference will be made to the intra-EU dynamics that lie behind the negotiating positions presented in international fora.

This chapter is divided into three parts. The first part situates the EU as a global actor, looking at the bases of the EU's authority to speak on behalf of its member states in global affairs. The second part looks at the negotiating position of the EU across a range of environmental problems, offering a series of explanations for why EU leadership is advanced in some areas, but much less forthcoming in others. In order to do this, both the internal features of EU policy-making and the actors involved in its formulation, as well as the multiple and often conflicting roles the EU performs in global affairs, are invoked to offer a fuller understanding of the Union's role in global environmental politics. The chapter concludes with some remarks about the prospects for EU leadership in global environmental affairs.

The EU in global affairs

Cooperation with third parties has been a feature of EU environmental policy ever since the first Environment Action

Programme in 1973. The European Union is now a contracting party or signatory to a wide range of international conventions in the environmental field. These include the agreements on global issues such as ozone depletion and the Basel Convention on Hazardous Wastes, as well as regional agreements such as the Barcelona Convention for the Protection of the Mediterranean (Haas, 1990). The EU is now a signatory to over 30 major multilateral environmental agreements on issues from fisheries and desertification to climate change and loss of biodiversity (Bretherton and Vogler, 1997). Yet despite this degree of involvement, there remains a sparsity of literature on the EU as an international actor in environmental politics (Sbragia, 1997). From a European studies perspective this is surprising when it is considered that looking at the EU as an actor on the world stage raises interesting questions about authority, subsidiarity and implementation within the EU.

This neglect is also problematic from an International Relations (IR) perspective given the increasing role of the EU in international diplomacy. In addition, it is notable that many of the difficulties which confront countries in their relations with one another over environmental issues in a global setting, can be seen in micro within the EU (Grubb, 1993). There are the very different levels of contribution to the problem, the wealth disparities and different attitudes towards environmental problems which make common policy difficult. There is a (relative) lack of institutions with the power to force actors to change their behaviour. All these characteristics are mirrored in international affairs, albeit in a more dramatic way. This makes the EU an important testing ground for policy initiatives: a laboratory in which to test the likelihood of international action on a range of threats to the global environment. The success of EU environmental policy therefore has global ramifications. The strategy the EU chooses to pursue will strongly affect the likelihood of other states cooperating with an agreement.

The salient position of the EU in global environmental affairs arises partly from the global economic status of the EU. In other words, the importance of the EU in the global economy forces it into a prominent role in the environmental arena. It becomes a key player in international environmental regimes because it is such an important trader and consumer of natural resources. As Bretherton and Vogler (1997, p. 2) note, 'By any standards the countries of the Union cast a long ecological shadow.'

The United Nations and its subsidiaries provide the principal fora for the negotiation of environmental agreements to which the EU is party. Whilst the current status of the European EU in the UN framework is based on the world of 1945 in which regional organisations did not exist and nation-states predominated, the UN has shown greater openness to EU participation since this time (Brinkhorst, 1994). The EU is an observer in most subsidiary organs of the UN such as UNEP (United Nations Environment Programme). It is often treated as an intergovernmental organisation and is accorded full membership of organisations such as the FAO (Food and Agriculture Organisation) because of its exclusive powers in the field of agriculture as well as de facto membership of GATT (General Agreement on Tariffs and Trade) derived from exclusive powers in the area of commercial policy. The right to participate in negotiations has to be agreed for each UN conference, however, and problems of shifting patterns of competence, lack of precedence and problems of monitoring and compliance collude to complicate third-party negotiations with the EU (Sbragia, 1997).

The legal basis for EU participation in international agreements has evolved from case law developed by the European Court of Justice (ECJ), most especially the ERTA case which permitted the expression of external authority in areas of internal competence (European Court of Justice, 1971). In other words, the EU's external powers can expand without the express approval of member states in the course of developing internal policies. The Single European Act (SEA) and the Maastricht Treaty have subsequently confirmed the competence of the EU to conclude international environmental agreements binding on the institutions of the EU and upon member states. The fact that external competence flows from internal decision-making powers means that member states are alert to the fact that agreeing to new internal legislation external may result in the extension of competence (Haigh, 1992). And yet the confirmation of the independent nature of EU external power by the Maastricht Treaty means increasingly that the EU does not have to rely upon the existence of internal measures to justify external competence. As far as third parties are concerned however, it is often the *perception* of the EU's authority that affords it credibility in international environmental negotiations (Bretherton and Vogler, 1997). So, for example, when the Presidency speaks at international meetings it is assumed that this is the view of the EU, whereas in practice only the Commission can perform this role.

The EU pursues foreign relations as if it were a sovereign state by entering into agreements with other nations. It signs environmental conventions alongside member states. Where the EU has competence (under a mixed agreement), the Commission negotiates on behalf of the Union in accordance with a mandate conferred upon it by the Council. The Commission works within a (usually) well-defined mandate agreed unanimously by the Council to negotiate on its behalf, whereafter the Commission has to return to the Council for approval for what it has negotiated. At this point the Council determines how agreements will be translated into member state and EU law. Where there is no unanimity among member states problems arise. This happened, for instance, at a meeting of signatories to the Convention on the Trade in Endangered Species (CITES) in Buenos Aires in 1985 where, in the absence of unanimity, all ten member states abstained. Where member states vote on their own (in areas where they retain jurisdiction), they are meant to respect the principle of EU solidarity and vote the same way. This has created problems in the past. In relation to the implementation of the Climate Convention, for example, the UK government indicated that it would ratify on a national basis if EU ratification was to be linked to the introduction of a carbon tax to which it was opposed (Haigh, 1996). The Council can also conclude agreements only after consulting with the European Parliament. The assent of the Parliament is required for agreements with important budgetary implications for the EU, or which affect the co-decision procedure. In the environmental area only general action programmes are covered by this procedure.

As the EU adopts a fuller role as a legislator in the traditionally sensitive and closed world of foreign affairs, there will be continuing turf battles over jurisdiction and authority over policy that will produce a perplexing situation of flux for the EU's partners in diplomacy. The dilemma the EU has to face is that the more it resembles a sovereign state, the more likely it becomes that parties with whom it is negotiating will insist on according it only one vote, therefore significantly diminishing the power of the EU bloc (Haigh, 1992). The intricacies of the competence issue are one of the problems highlighted in the case studies which look at the EU's involvement in international efforts to address global environmental problems.

Theorising the EU in global environmental affairs

The fact that it is so difficult to pigeonhole the EU as a political and legal entity helps to explain the underdevelopment of theoretical perspectives on the EU in the global politics of the environment. There is a schizophrenia at work where the EU simultaneously tries to perform two roles, that of representative of its member states and that of distinct actor with leadership ambitions. It is sometimes more helpful to think of the EU as an international organisation, on other occasions a state, and on other occasions still, it fits into neither category depending on the division of competencies among its constituent parts (Bretherton and Vogler, 1997). Whilst performing many quasi-state functions, the Union does not bear the hallmarks of statehood in the traditional and narrow way in which scholars of IR would choose to define it, a fact which allows them to exclude it as a significant unit of analysis.

While there has been some attempt to employ IR approaches to understand the integration of states within Europe (Breckinridge, 1997; Hurrell and Menon, 1996), the policy-making processes of the European EU (Haas, 1998) as well to explore the channels through which global forces penetrate the EU, there has been little attempt to explore the use of IR theory for accounting for the role of the EU in global affairs. The role of the EU in international environmental politics has not escaped this neglect.

Regime theory might expect to tell us a great deal about the role of the EU in global environmental affairs. It may be insightful to think of the EU as a regime given that the regime definition most frequently cited is so broad as to certainly include the EU where 'norms, rules and decision-making procedures in a given area of international relations' (Krasner, 1983, p. 2) are said to be in existence. This sort of an approach would enable us to think about the relationship between the institutions of the EU and the member states. It may fruitfully account for the inter-state bargaining that lies behind the formation and development of EU-wide policies on the environment given that the positions the EU projects in international negotiations are clearly themselves the product of interest mediation and contractual bargaining overseen by institutions.

Yet for the most part, regime theory, the most commonly employed theoretical paradigm in the study of international environmental politics (Smith, 1993), is silent on the role of regional economic organisations such as the EU. This neglect is either deliberate, unfortunate or both. Perhaps deliberate because the involvement of the EU in the development of these international regimes to protect the environment unsettles a number of assumptions which regime theory employs to account for outcomes in global environmental politics.

The little mention that the EU does receive in his literature focuses upon how the EU affects the prospects of regime-building and how it may complicate the path of international cooperation (Haas, Keohane and Levy, 1993) consistent with the 'problem-solving' approach within which regime theory is located (Cox, 1987). By signing up to agreements on behalf of its member states, the EU may extend the scope of a regime by ratcheting up the obligations of states that may otherwise have adopted lower standards. The drive of the 'push' states (Sbragia, 1996) in international environmental negotiations sets the pace for the EU as a whole with the effect that 'pull' states are drawn into commitments beyond what they would have accepted of their own accord. Often, however, the 'convoy' analogy (Bretherton and Vogler, 1997) more accurately captures the process, whereby action is delayed by the slowest part of the train. We saw this effect during the ozone negotiations, where, despite the efforts of Denmark and Germany to push things forward, the obstructive tactics of France and the UK were able to ensure that on many occasions the EU was 'condemned to immobility' (Jachtenfuchs, 1990, p. 265). Nevertheless, by coordinating the position of (currently) 15 nations in environmental negotiations, the Commission reduces the complexity of negotiations (Bretherton and Vogler, 1997) and eases pressures upon international organisations to perform that role. Monitoring and verification are also easier with smaller numbers (Haas, 1998). Where international secretariats often lack resources and authority to monitor and oversee the implementation of commitments, the EU can perform this function in relation to its own member states.

Realist and liberal institutionalist concerns about the potential for free-riders have also clearly played a part in the EU carbon tax debate (see Chapter 5) and help us to understand efforts to bring India and China within future global climate accords. Approaches informed by game theory would also help to explain the leadership role of the EU as an attempt to initiate cooperation conditional on the involvement of other parties. Hence the declaration of a greenhouse gas abatement target as early as 1990 was intended as a first move in the 'nice, reciprocate, retaliate' strategy that Axelrod (1990) suggested is the key to cooperation. Paterson (1996, p. 105) notes, for instance, that 'The announcement of the EU target in October 1990 was explicitly designed to influence the outcome of the Second World Climate Conference and to precipitate international negotiations'.

In general, however, IR perspectives are prone to overlook the importance of intra-country dynamics to the formation of positions in international negotiations; a factor which severely circumscribes their applicability to EU decision-making processes. Nevertheless, Putnam's (1988) two-level game would help to address member-state–institution interactions. In the ozone case, as with climate too, it could be argued a mixture of 'domestic' and international pressures best explain the role of the EU in establishing and maintaining the regime in question. The use of a two-level game in this context would involve treating the EU as one unit however. To do so would distort the picture to such a degree that four-level games may provide a better framework (Matláry, 1997). The four games being played simultaneously are: one between member states and the EU; between the EU institutions in their internal power struggle; among the directorates and finally between the various directorates and interest groups (Matláry, 1997, p. 146).

An approach which gets inside the 'black box' of the EU and

its member states is likely to prove more useful than an intergovernmental approach or even a two-level model which would exaggerate the extent to which the EU can accurately be represented as a single political unit. A four-level game would allow us to capture, for instance, the role of industries, key players within the Commission and Parliament, internal EU policy-making battles, as well as interactions with third parties. This problematises notions of unitary actors, rational decision-making and static national interests derived from an actor's place in the international system. The dynamics of EU policy-making require a less state-centric transnationalist framework. This is especially appropriate for understanding the role of EU in global affairs, where there are multiple national and supranational interests (such as promoting integration and extending the competence of the EU), where unitary action is particularly difficult, and decision-making takes place across a number of 'levels'. An approach which examines member states' attitudes and interests in the EU and links these with developments in international politics, as Bulmer (1983, p. 363) argues, 'corresponds most closely to the transnationalist approach of international relations theories'. The way in which during international negotiations the Commission is forced to continually report back to the Council on its mandate and to agree a proposal under discussion suggest the importance of a multitiered approach. When negotiating an international environmental agreement policy-makers have to bear in mind the repercussions for their place in Europe as well as the implications of the deal for their own country, and the place of both in a global context. They play a number of games on different levels on a simultaneous basis.

The role of the EU in global environmental politics is, therefore, subversive of many of the assumptions implicit in traditional International Relations (IR) approaches. The EU problematises regime analysis because it raises difficult issues about unitary actor behaviour where the EU has to develop perceptions of mutual interest among its own members before it can contemplate the striking of accords with third parties. The (particular) importance of interest groups lobbies, which are reflected by regime analysis, in the setting of the EU's position on environmental questions, also makes it harder to determine whether a position has been brought about of inter-state interaction or because of interest group pressure. The pluralist assumptions underlying regime analysis are also rendered problematic by the structural advantages afforded to industry lobbies in the EU derived from their central position in enabling the process of economic integration and facilitating the growth objectives of the EU (see Chapter 2). The transnational scope of the lobbies that operate within and beyond the EU complicate the convenient delineations between domestic and international politics. The complex structure of the EU challenges the very usefulness of the terms 'domestic' and 'international' as they are applied in regime analysis (Liberatore, 1997). The dense and multitiered network of governance structures which operate within the EU are not accurately described by the binary oppositional language of domestic/international. The EU occupies a space between the national and the international worlds of environmental politics. In many ways it has to defend its legitimate function in environmental affairs both to member states who want to see evidence of the value-added from the participation of the EU institutions and to other international negotiators who are often sceptical and bemused by the rights and responsibilities of the EU in negotiating fora where nation-states have traditionally been the only legitimate actors.

The preoccupation of regime perspectives with a very narrowly defined issue-area also serves to obscure the multifaceted global personality of the EU. No account is likely to be able to explain fully EU cooperation in international arrangements to protect the environment unless it is sensitive to the economic context in which it takes place. The criticism that regime approaches fail to locate state actors in a global economic context necessary to understanding their role in international environmental cooperation (Newell and Paterson, 1998) applies even more strongly to the study of the EU where the EU's participation in international environmental regimes is often explicitly framed in terms of the goals and demands of European economic integration.

With the EU we clearly have a regime within a regime, where models of multi-level governance used to explain the policy process within Europe (Lenschow, 1997; Grande, 1996; Weale, 1996) may be extended to incorporate the international dimension. State-centric theories of IR are, at this stage, ill-equipped to capture the dynamics of regional economic organisations within international institutions. A political economy of the EU in the global system is likely to be more sensitive to the dynamic nature of the EU's role in global affairs as well as the increasingly important role of transnational corporations (TNCs) in establishing and maintaining connections across issue areas and institutional fora. A transnationalist approach (Risse-Kappen, 1995), freed of some its pluralist underpinnings, may offer such a framework. It would be well placed to capture the interdependencies among and between states, non-state actors, regional organisations and international institutions; to explore the fragility of domestic-international separations and to explode the myth of unitary actor models which obscure the importance of bureaucratic politics.

Viewed from this perspective, international environmental negotiations become a site of contest between transnational networks of environment departments from government and regional economic organisations working together with NGOs and sympathetic international organisations (such as UNEP), set against networks including Trade and Industry departments, business lobbies and international organisations which promote the interests of industry (such as UNIDO (United Nations Industrial Development Organisation)). The networks operate horizontally and vertically and across national, regional and international levels involving state and non-state actors alike in strategic alliances formed on particular issues. This helps us to get beyond the 'billiard-ball' model of IR which takes as given the clashes between autonomous and unified political units and expresses more accurately the nature of the EU's insertion into world politics.

Case studies

Ozone: The first global challenge

The evolution of the regime designed to limit the release into the atmosphere of ozone-depleting chemicals is in many ways a story of EU–US relations (Rowlands, 1995, 1997; Jachtenfuchs, 1990; Benedick, 1991). The key turning points in the progress of the negotiations from a framework convention at Vienna through to legally binding protocol commitments at Montreal, London and Copenhagen reflect changes in the negotiating position of the EU and the US. The centrality of the EU in the creation of the ozone problem, as with many other global environmental problems, meant that it was forced centre stage.

The history of ozone politics can be dated back to 1977 when the 'can ban' instituted in the US (a restriction upon the use of Cchlorofluorocarbons (CFCs) as propellants in non-essential aerosol sprays) put the US in a position to push for a global ban on CFCs. Negotiations moved very slowly at first against strong European opposition to cuts in CFCs, despite a Council decision in March 1980 limiting the use of CFCs, reacting to American pressure and growing public concern over the ozone issue (Jachtenfuchs, 1990). The advocates of controls (the US, Canada, the Nordic states, Austria and Switzerland), came together in 1984 to form the 'Toronto group'. The EU initially argued that no controls were required, but eventually conceded that a production capacity cap may be necessary and proposed a draft protocol that mirrored their own 1980 measures. The proposed 30 per cent reduction was easily achievable because use was already declining (Haigh, 1989; Rowlands, 1998) and essentially served to fix the status quo (Jachtenfuchs, 1990).

The deadlock that ensued between the EU and the Toronto group ensured that only a framework convention could be agreed at Vienna. This pledged cooperation in research and monitoring and encouragement of information-sharing. At the March 1986 meeting of the EU Council of Ministers, the EU moved to a position of a 20 per cent CFC production cut, partly prompted by the threat of unilateral action by the US to impose trade sanctions against the EU (Rowlands, 1997, p. 7). The Montreal Protocol subsequently agreed in September 1987 called for cuts of 50 per cent from 1986 levels of production and consumption of the five principal CFCs by 1999. The figure of a 50 per cent cut was an arbitrary one designed as a compromise position between the EU's proposed freeze and the US's proposal for a 95 per cent cut. The Protocol included a time-lag for the implementation of the Protocol by less developed states, restrictions on trade with non-parties and an ozone fund for technology transfer. This latter aspect of the agreement is particularly important for the EU for, as Jachtenfuchs (1990, p. 272) notes, 'The success of the EU's environmental diplomacy in this important field will to a large extent depend on how far it is able to provide technical and financial assistance to developing countries'.

As a regional economic integration organisation, the EU was permitted to meet consumption limits (but not production limits because of US objections) jointly rather than country by country. This was intended to allow some transfers of national CFC production quotas among EU states to allow commercial producers in Europe to rationalize production facilities cost-effectively. Despite this concession, some European participants in the Protocol process felt they were 'bullied' into an agreement favourable to US industry, dubbing the Montreal agreement 'The DuPont Protocol' (Parsons, 1993, p. 61). Nevertheless, on 14 October 1988 the Council adopted a regulation, transforming every detail of the Protocol into EU legislation. The regulation came into force immediately in order to underline the importance of the issue and to avoid trade distortions that might arise from non-simultaneous application of the proposed legislation (Jachtenfuchs, 1990, p. 269).

At the March meeting of the EU Environment Council of 1989, the UK finally joined the rest of the EU in agreeing to phase-out all CFCs 'as soon as possible but not later than 2000' (Parsons, 1993, p. 47) and France submitted to external pressure to drop its intransigent position. The London meeting of the parties in June 1990 was therefore able to agree that all fully halogenated CFCs would be phased-out by the year 2000, with interim reductions of 85 per cent in 1997 and 50 per cent in 1995. Some member states have gone beyond the commitments required by the international agreements, however. Germany, for example, has put in place legislation requiring that CFCs be eliminated by 1993, halons by 1996, HCFC 22 by 2000 and CT (carbon tetrachloride) and MC (methyl chloroform) by 1992 (Parsons, 1993). The EU as a whole met its 1996 target for phasing out CFCs and halons and claims to be on course to meet the targets for HCFCs under the Montreal Protocol (DGXI 1998).

On another level, behind the inter-state diplomacy of the negotiations, the story is essentially one of the competing positions of the chemical companies, most especially, ICI (in the UK), Du Pont (in the US) and Atochem (in France). Industry representatives served officially on European national delegations throughout the process (Parsons, 1993; Litfin, 1994). EU industrialists 'believed that American companies had endorsed CFC controls in order to enter the profitable EU export markets with substitute products that they had secretly developed' (Benedick, 1991, p. 123). Movement in the EU's negotiating position during the summer of 1987 stemmed from a relaxation of the British attitude following ICI's development of substitute chemicals. This supports the argument both of Rowlands (1997), who shows that competitiveness concerns have been at the heart of EU decision-making throughout the history of ozone diplomacy, and Benedick (1991, p. 68) who maintains that the EU 'followed the industry line and reflected the views of France, Italy and the United Kingdom'.

The importance of these commercial considerations is manifested in the continuing efforts to agree cuts in HFCs and HCFCs (perceived to be the best alternative to CFCs). The EU has found it difficult to come to a common position on reducing the production and consumption of these chemicals because substitutes are not yet readily available. Attempts by Germany and Denmark to take the EU position beyond the terms of the Copenhagen agreement met with an intervention by Commission President Jacques Delors pursuing objections made by the French company Atochem (Rowlands, 1997). Hesitancy can also be attributed to the fact that some European producers want to create export markets for HCFCs in the less developed 'south'. The differing commercial interests in relation to the ozone issue highlight the difficulty the EU faces in trying to formulate common policy positions in international environmental negotiations.

It is important to note briefly a range of other factors which meant that the EU was eventually able to adopt a more progressive position on this issue. As has already been noted, the EU was under considerable pressure from the US (backed by its chemical companies) to agree phase-outs. The pressure from the US tapped into an acutely felt need by the EU to demonstrate a responsible common position on environmental issues (Jachtenfuchs, 1990). Related to this was the popular concern about the 'hole in the pole' (following work by the British Antarctic Survey in 1985 which alerted the world to the extent of the depletion of the ozone layer); prompting a scare which blew up in 1985 about the immediate effects of ozone depletion on human health, most notably in the form of increased exposure to cancer-causing ultra-violet rays, stimulating popular awareness manifested in widespread consumer boycotts in both the UK and the US directed against chemical companies continuing to manufacture CFCs. The European Environment Bureau also organised symposia and disseminated key scientific findings at critical moments in the negotiating process designed to sway public and governmental opinion (Seaver, 1997, p. 56).

Germany was also pushing for more stringent controls and legislated controls at the national level that went far beyond the requirements of the international agreements. During the negotiations for the Montreal Protocol, Germany pushed for an aerosol ban and a 50 per cent overall reduction under pressure from a vocal green lobby and a Green Party rising in popularity. German industrialists were also anxious to see a Europe-wide solution, on the grounds that it would probably be weaker than regulations implemented at the national level and harmonisation of standards would minimise the costs of having to manufacture products to different criteria across the EU. The Dutch and Belgians were also part of this more proactive coalition within the EU. The shift away from British dominance in the EU towards a greater German role was significant in making the EU a more cooperative partner in the international negotiations (Seaver, 1997; Benedick, 1991). By mid-1987 West Germany had 'essentially replaced the UK as the primary influence on EU ozone policy' (Benedick, 1991, p. 39).

Ozone depletion was one of the first global environmental issues to produce a coordinated international response. Despite ongoing weakness in the ozone regime it is considered to be one of the few tangible successes of international environmental diplomacy given that governments took action before definite proof of environmental damage had occurred.

The big 'Green' summit: EU at UNCED

As with its standing with UN organs, the status of the EU at conferences held under UN auspices is governed by a General Assembly resolution of 1974 granting it status as a non-voting observer. However, based on a recommendation of the Preparatory Committee of UNCED (United Nations Conference on

Environment and Development), the General Assembly decided by a special decision of 13 April 1992 to grant the EU upon its request 'full participant status' at the Rio negotiations (Brinkhorst, 1994). Full EU participation at the UNCED meeting was approved by the Council of Ministers 'on equal terms with member states' (Sbragia, 1997, p. 26). This confers on the EU the right to representation in committees and working groups of the conference, the right to speak and reply and to submit proposals and amendments. The EU did not, however, have the right to vote or to submit *procedural* motions. This special position was granted on an ad hoc basis and is not necessarily thought to constitute a precedent for the future. The status and entitlements of the EU have to be negotiated at each series of negotiations on the environment.

On this occasion, the EU was able to claim this status because many of the items listed on the UNCED agenda were already covered by legal acts adopted by the EU. This applied in relation to the protection of habitats as well as to 'non-environmental areas' such as trade, transport and development which nonetheless have important environmental impacts. During the course of the conference, the Commission was given the responsibility of presenting the EU position, negotiating on its behalf and expressing its views on all questions falling within the EU's exclusive powers. The EU was left to inform the UNCED Secretariat before consideration of each item by the conference, whether the EU was competent and who would speak for it.

The EU came to the UNCED debates with an outlook supportive of a substantive outcome. During the preparatory process, the EU had played a mediating role in attempting to bridge the gap between developing countries on the one hand and the US, in particular, on the other. This was particularly so with the Climate Convention, where the US was fundamentally opposed to the inclusion of binding commitments to reduce greenhouse gases in the final convention and many less developed countries were resolutely opposed to being forced to adopt any commitments towards alleviating a problem to which they had contributed very little. The EU had shown itself to be more sensitive than many of its OECD partners to the argument that it would be unfair to impose further obligations on less developed states until industrialised parties had taken substantive action (Grubb, 1995).

In relation to both the Climate and the Biodiversity conventions, the EU performed a leadership role and in relation to other outcomes of the conference, such as the Rio declaration on Environment and Development and the Forest principles, the EU's role is described as that of a 'mediator' (Brinkhorst, 1994). This section briefly reviews the involvement of the EU in the Climate Convention, and the negotiation of the Forests declaration and refers briefly to the EU's perspective on the other issues discussed at Rio.

(i) Climate change

Discussion of the involvement of the EU in the international climate change regime will be kept brief given that Chapter 5 is devoted to the intra-EU politics of the issue. Nevertheless the global politics of the issue, in as far as they can be distinguished from the policy process within the EU, shed light on different dimensions of the role of the EU in external affairs.

The climate change issue hit the international political agenda in the late 1980s amid public anxiety, fuelled by the media, that the series of freak weather events being experienced around the time were evidence of future catastrophes at the hands of climate change (Ungar, 1992). A number of scientific reports, most prominently those of the Intergovernmental Panel on Climate Change (IPCC), established climate change as a serious concern and endorsed political action by way of response (Houghton et al., 1990). It was not long before the United Nations set negotiations in train towards a global instrument to address the issue. Relatively distinct negotiating blocks rapidly formed (Paterson and Grubb, 1992). At one end of the spectrum was the Alliance of Small Island States (AOSIS), a group of states most vulnerable to the projected impact of climate change (most especially sea-level rise), calling for immediate international action to contain the rate of greenhouse gas emissions growth. At the other end of the spectrum the Oil Producing and Exporting Countries (OPEC) group consolidated itself around a resistance to action on climate change given the threat posed to their economies implied by a shift away from fossil-fuel consumption. In between this polarity was scattered a diversity of positions from the United States (US) towards the anti-action line of the OPEC states and the bulk of G77 least industrialised countries showing little initial interest in the issue other than to underline the contribution of Northern hemisphere countries to the problem, and compare this with their own comparably minor contribution. The European position surfaced at the more pro-action end of the spectrum.

The EU is forced to play a key part in the international negotiations on climate change because of the percentage contribution it makes to global emissions of greenhouse gases which currently stands at 16 per cent. The EU has been regarded as a leader on this issue by virtue of the intransigence of most other industrialised countries and in the first instance by declaring a commitment to return its joint CO_2 emissions to 1990 levels by the year 2000. In the early negotiations on climate change there was little EU competence and the Commission was not a formal participant at the Intergovernmental Negotiating Committee (INC) meetings. Because of the number of DGs involved in the interservice committee on the climate change issue, internal deliberations were often difficult and extensive (Sbragia, 1996). Perhaps as a result of this, Bretherton and Vogler (1997) argue that the Commission has been relatively inactive on the climate issue, preferring instead to rely upon lead countries (Netherlands, Germany, Denmark) to provide inputs.

In October 1990, the EU and its member states decided to pursue a stabilisation target based on 1990 levels (Vellinga and Grubb, 1993). This commitment is in fact a 'clearer and stronger statement' (Macrory and Hession, 1996, p. 140) than that contained in the Convention. EU member states were among the first to adopt targets for limiting CO_2 emissions and to urge the international community to negotiate a binding Convention including emission constraints.

It was anxiety on the part of the EU that a Climate Convention would be meaningless without the signature of US President Bush (given that the US is the single largest contributor to the problem) that prompted a spate of shuttle diplomacy between London and Washington on the part of the UK's then Environment Minister, Michael Howard, trying to convince the President that the Convention was sufficiently flexible for the US to be able to sign it without inflicting any damage upon the US economy in an election year. As with the ozone issue, the run-up to the agreement of the Framework Convention on Climate Change was therefore a story of EU–US diplomacy whereby the EU (and individual member state governments working on their own) was forced to lobby the US to stay on board whilst lowering its own expectations about what the Convention would contain. Michael Howard, acting with the support of EU environment ministers (but without a formal mandate from the Council), managed to bring about agreement on the basis of Article 4(2) of the Convention which spells out the commitments for industrialised countries (Haigh, 1992). The wording in the Convention both on the need for industrialised country parties to stabilise their emissions of CO_2 and for the Convention to be guided by a precautionary principle reflected formulations by EU ministers (Grubb, 1993, pp. 24–5). In the end binding targets were not contained in the Convention and parties are required only to 'aim' to stabilise their emissions of greenhouse gases at a level that will not permit dangerous interference with the climate system (a level which is not specified in the Convention).

The EU interpreted this to imply a return levels of CO_2 emissions to their 1990 levels by the year 2000; a commitment which it intended to implement either individually or on a collective basis. The EU target was adopted with the expectation that less-developed regions (such as Spain, Greece and Portugal) would be permitted to increase their emissions. In effect, therefore, higher-emitting Northern European states have accepted two commitments; one directly to the Convention (stabilisation) and the other to the EU as signatory in its own right, which may require them to achieve emissions *reductions* to accommodate the expanded emissions of their (mostly) Southern counterparts through the joint implementation of commitments (Grubb, 1993).

It is tempting to think that the highs and lows of international climate politics broadly reflect (or at least coincide with) the waxing and waning of EU leadership on the issue. The two most significant agreements to date (the convention itself and the Kyoto Protocol) both came at a time of US intransigence on the issue and EU assertiveness aimed at drawing the US into a compromise accord. The agreement of the Kyoto Protocol at the end of 1997, in particular, highlights the ongoing centrality of EU–US relations to the success of international climate initiatives. John

Prescott, the UK Secretary of State for the Environment at the time, was instrumental in ensuring that the US signed up to the Protocol. In order to achieve the US's consent however, the EU was forced to abandon its aim of encouraging all annex 1 (most industrialised parties) to agree to a 15 per cent reduction in CO_2 emissions by the year 2010.

Instead, the protocol allocates an individual emission reduction or stabilisation obligation to individual signatories (Newell, 1998). The EU came away from the agreement with a CO_2 reduction target of 8 per cent, of its 1990 levels to be achieved over the period 2008–12. The proposal to differentiate commitments to cut emissions of CO_2 within an overall EU-wide bubble ran into trouble however, particularly since the EU was against the use of the differentiation principle by other states. Complications arise from the fact that 'In the absence of a clearly defined area of exclusive EU competence for climate change and in the absence of a clear obligation detailing specific action, it is extremely difficult to isolate EU and Member state obligations' (Macrory and Hession, 1996, p. 114). Aside from the ongoing internal deliberations over target allocation, the other pressing issue is where the ceiling should be set upon the use of the Kyoto mechanisms (the Clean Development Mechanism, Joint implementation and permit trading) to meet emissions reductions obligations (see Chapter 5).

The EU has acknowledged both that the science gives a clear indication that further action is necessary and that significant greenhouse gas reductions are technically possible and economically feasible. And yet by the Commission's own admission, the EU-wide stabilisation target is likely to be overshot by a margin of between five and eight per cent (SEC (95) 288 Final). In 1996 the Commission Second Evaluation of Member States progress in meeting their commitments under the Climate Convention, reported that the EU 'is not in a position to claim that the adopted policies will be sufficient to meet the agreed targets and certainly not to ensure reductions in CO_2 emissions after the year 2000' (COM (96) 91 Final). It is ironic then that the EU is calling for limitation and reduction measures beyond the year 2000 (SEC (95) 288 Final), when it has experienced enough difficulty getting within sight of the goal of stabilisation at 1990 levels. There is an evident tension that will have to be confronted between the rhetoric of the EU, which still makes it the most progressive Annex 1 negotiating bloc, and the internal failure to find policies to meet the international leadership commitments.

It is an indictment of the state of international climate policy that despite numerous problems with the implementation of its commitments within Europe and the failure of a number of the key tenets of EU climate policy, the EU is still regarded as the leader on this issue. Kyoto did nothing to challenge this view, where compared with the increasingly intransigent Australian position under the Howard administration, a US delegation straightjacketed by a Republican dominated Congress hostile to action on climate change and a failure of Japanese leadership on the issue (despite being the host nation for the conference), the EU emerged triumphant among the most industrialised countries with the most far-reaching target (Newell, 1998).

(ii) Forests

The policy of the EU towards the conservation of forests is perhaps the most underdeveloped of the issue areas discussed in this chapter. The EU's concern over forests is often related to the climate change issue and has been a central component of the EU's strategy for combating global warming (Jachtenfuchs, 1996). Two of the three pillars of the Fitzsimmons Report on climate change in 1986 relate to afforestation and the development of policy measures to slow the rate of tropical deforestation and the fourth environmental action plan (1987–1992) devotes a large amount of attention to the protection of tropical forests.

At the international level agreement with third parties on this issue has been hard to come by. The non-binding document on forest principles that emerged from the Rio conference was a far cry from the Global Forests Convention (GFC) that some had hoped for. The issue was hostage to a series of ongoing (principally North–South) debates over sovereignty versus the collective ownership of the forests under the rubric of 'common heritage of mankind' [sic] (Humphreys, 1996). It appeared there was little the EU could do in the context of these broader, deeply rooted and ongoing tensions. Nevertheless, Professor Toepfer, working with a broad mandate from the Council, helped to avoid total breakdown on the issue at the final negotiating session on the forest principles document. Humphreys attributes the successful negotiation of this agreement to his involvement. He notes (1993, p. 51) 'Agreement on the statement of forest principles was reached after Klaus Topfer, the German Environment Minister, assumed responsibility for ministerial level negotiations.'

It was the EU, moreover, that shifted position at the Geneva preparatory meeting in April 1991 to advocate a compromise solution aimed at bridging the North–South conflict: a general declaration on forests including guidance on the text of a future Global Forests Convention, to be negotiated after Rio (Humphreys, 1997, p. 91). The EU issued a statement that any declaration opened for signature at Rio should contain 'procedures, including a timetable for the negotiation of a Convention on Forests' (quoted in Humphreys, 1996a, p. 99). Humphreys describes the effort made by the EU to break the G77 veto against a GFC as 'clumsy' (1996a, p. 161), however. In a classic example of diplomatic issue-linkage, when a group of African states called for a Desertification Convention, some EU governments sought to link this with the negotiation of a GFC post-UNCED. The intention was to split the G77's unified stance against the GFC by luring the African governments into the pro-GFC camp with promises of EU support for a Desertification Convention (ibid.). The EU failed to deliver on what could have been an important diplomatic breakthrough in the negotiations after the tactic was not endorsed by all EU members and was further undermined by US support for a Desertification Convention. It subsequently dropped its insistence that the desertification and forests issue be linked.

Moreover, the EU has encountered a number of difficulties post-Rio in taking action on forests, exacerbated by the fact that the EU has no competence in this area even though DGIV and DGVIII have polices on certification and development aspects of forestry (Bretherton and Vogler, 1997). The EU nevertheless adopted a forest strategy in 1992 to support afforestation through national plans. The EU contributes between 50 and 70 per cent to the costs of the project. It has also emphasised sustainable forest management in its agreements with developing countries and added a protocol on Sustainable Management of Forest Resources when the Lomé IV Convention was revised in 1995. Tropical forests have become central to the EU's external environmental assistance efforts bringing together commitments under Agenda 21, the Rio declaration and Forest statement as well as climate change, biodiversity and desertification (Robins, 1998). As part of the International Tropical Timber Agreement between 1992 and 1996, the EU also allocated ECU50 million to promote the conservation of tropical forests (DGXI, 1998). Within the ITTO (International Tropical Timber Organisation), the body which oversees the above agreement, the EU also holds the second largest share of consumer country votes after Japan (Humphreys, 1996) and is well placed therefore to influence the future course of the global forests debate. At the national level there have also been efforts towards certification of sustainably produced timber in countries such as Belgium, Sweden and the UK (Saint-Laurent, 1997).

(iii) Other issues at UNCED

In general terms, as Brinkhorst (1994) argues, the EU seems to have enjoyed most success at Rio in advancing those initiatives which overlap with a body of pre-existing internal programmes and least success in areas such as finance, reflective of the comparative underdevelopment of EU Development Aid policy. At Rio, governments demonstrated a collective unwillingness to pledge new financial resources to support environmental programmes. The issue of 'additionality' featured prominently at UNCED where there was some dispute over whether funds, additional to existing development budgets, would be made available for the implementation of the environmental commitments of less developed countries. It came up in the light of the EU's offer of a 3 billion ECU contribution to the 'quickstart' fund for the implementation of Agenda 21, where it was unclear whether new and additional resources would be found for this (Brinkhorst, 1994; Hull, 1994). There was no common EU position on financial questions because of the limited capacity of the institutions of the EU to speak on financial questions. Five years after the Earth Summit, the EU has failed to deliver on its $4 billion aid commitment towards implementing sustainable development, pledged for the implementation of Agenda 21 in LDCs (Robins, 1998, p. 190).

Many regard Agenda 21 as the most significant document at Rio in terms of its breadth of coverage. As a result, the EU's fifth

Environmental Action Plan is guided by the underlying principles of Agenda 21 and the Rio declaration (Dent, 1997). Agenda 21 implementation reports from the EU to the CSD have emphasised that the 5th EAP 'has been chosen as the community's main vehicle for implementing Agenda 21 and other agreements made at UNCED' (quoted in Wilkinson 1997, p. 158). And yet only seven member states were found to have prepared sustainable development strategies in accordance with Agenda 21 when the Commission reviewed its interim review of the 5th EAP (Dent, 1997). As is the case with most international environmental accords, they only take effect once implemented at the national or subnational level, where the EU has less power to ensure that commitments are enforced. Some areas targeted by Agenda 21 primarily fall within the competence of member states; in other areas the EU takes a lead.

The UNCED process of ongoing negotiations, periodic review of progress and monitoring requirements, has locked the EU into a continuing dialogue with other states about the most effective way to address a range of global threats to the environment. Its record to date is a mixture of leadership coups and public embarrassments. Internal wrangles over competence, and pressure from third parties will undoubtedly collude to shape the future participation of the EU in global affairs.

The EU in implicit global environmental politics

Explanations of EU external environmental policy must also take account of what Conca (1993) refers to as implicit environmental politics; areas of activity that are not officially ascribed the term 'environmental' but which nevertheless have a significant impact upon the effectiveness of environmental policy. In this respect the activities of the EU in 'other' areas of global politics will have a bearing on its ability to handle global environmental problems. The numerous international institutions in which the EU performs a leading role mean that it is a significant player in areas as diverse as trade, aid agriculture and transport, activities which all have enormous ecological significance.

There is some indication that the EU recognises and is acting upon these connections. Robins (1998) shows that EU has reviewed its aid programmes in the light of their environmental effectiveness and impact. This has involved the incorporation of project cycle management provisions into the Lomé and other aid programmes. The Lomé IV convention, signed in December 1989, stated a commitment to screen all aid-funded projects with an environmental impact assessment and lists environmental protection and conservation as basic objectives (Hull 1994). Nevertheless parliamentarians and environment and development NGOs have been critical of the environmental and social impact of controversial aid initiatives such as the Carajes steel programme in Brazil, the Kibale forest scheme in Uganda and the Risonpaln oil plantation in Nigeria (Robins, 1998, p. 192).

An attempt has also been made to shape the environmental policy of third parties through aid to the transition countries of central and Eastern Europe (CEE). In this regard, as Liberatore notes (1997, p. 197), 'Negotiations regarding the accession of new member states offer the EU a particularly valuable occasion for agenda-setting.' A number of programmes have been put in place to this effect. Among the most prominent is the PHARE programme (Poland and Hungary: Action for Restructuring the Economy) initiated in 1989 at a G7 summit. PHARE is aimed at capacity building to deal with the most urgent environmental problems in the short term and the harmonisation of Europe-wide environmental standards in the longer term. Through PHARE the EU committed 337 million ECU to its environmental programming in Eastern Europe from 1990 to 1994 (Connolly et al., 1996, p. 287) in the form of a grant, making the EU the largest grant-based donor of environmental assistance.

Criticisms of the programme have focused on the fact that at present only 9 per cent of the PHARE budget is devoted to environment and nuclear safety. Similarly, of the 1870 million ECU that TACIS has provided for more than 2000 projects in the newly independent states, only 256.3 million ECU of this total has been allocated to top priority nuclear safety and environmental projects (Liberatore, 1997, p. 202). The nature of aid environmental aid packages offered by the union repeat many of the same errors as other aspects of its environmental policy, where projects are funded which fundamentally clash with the objective of protection the European environment. As Baker notes (1996, p. 153), 'it continuously gives priority to economic reform, even if these measures bring negative environmental consequences'. EU funding of transport infrastructure in CEE is primarily focused upon road building. One project, for example, is aimed at rehabilitating 800 km of roads and motorways linking Romania with Greece and those linking the former Yugoslavia with Turkey and the Black Sea and completion of parts of the trans-European motorway (Baker, 1996, p. 164).

Connolly and List (1996) suggest two reasons for the failure of the EU to use aid programmes to impose further environmental conditionalities. Firstly, PHARE and TACIS lack adequate financial inducements to pursue such a strategy. Secondly, there are divisions among member states about how highly they value the closure of nuclear plants in CEE over the commercial benefits of enhancing the investment opportunities of their nuclear companies. A third may be that if the EU were to impose conditionalities on its aid packages, other bilateral funding arrangements, where conditions are not imposed, would become more attractive and from which European industries could not be guaranteed the same slice of the action. And finally, those in charge of the programmes have tended to take the view that determination of key areas such as national energy policy is the responsibility of the recipient country.

The extent to which the EU will be willing to pursue 'green' development through its aid policy in the future is unclear. As Baker (1996) notes, economically weaker 'cohesion' countries within the EU may start to resent the scale of resources being ploughed into environmental reform in CEE when they are faced with many pressing problems themselves which are receiving less attention as a result of the budgetary demands of TACIS and PHARE. It is also possible that aid to CEE will face direct competition from the requirements upon the EU to deliver financial aid to LDCs (lesser developed countries) through its international obligations under the climate and other international conventions. The environmental share of the PHARE budget has already been allowed to drop from 20 per cent in 1990 to 6 per cent in 1992 (Baumgartl, 1997, p. 173). This suggests that, for the time being at least, calls for environmental 'Marshall Aid' from the EU to CEE will remain pipe-dreams, even if countries such as Austria are prepared to commit their share of the financial resources towards such a programme (Liberatore, 1997). Another limit to the EU using development policy as a vehicle for 'ratcheting up' environmental standards is that many areas member states 'jealously preserve traditional bilateral aid arrangements with their attendant advantages for pursuing national security interests and promoting national exports' (Golub, 1998, p. 26). A further shift of decision-making on aid policy to the European level may therefore be resisted.

Another key area of international policy with important environmental repercussions is trade. Through its influence within the WTO (World Trade Organisation), the EU could be a critical agenda-setter in the trade–environment debate. The Commission (which in practice represents the EU at the WTO) could call for a strengthening of the CTE (Committee on Trade and the Environment), or push for a legitimation of process-related product discrimination. And yet at present the EU's position on the use of trade measures for environmental protection seems at best schizophrenic, at worst blatantly self-interested. In relation to the infamous Dolphin Tuna case of 1988 (eventually concluded 1991), the EU accused the US of violating GATT rules by pursuing environmental protection beyond its own borders. On other occasions, as Golub notes, it has endorsed the need for governments to incorporate extraterritorial environmental considerations into their trade policies, such as in relation to PIC regulations for the export of pesticides. Similarly in relation to TREMS (trade-related environment measures), environmental clauses attached to the EU's GSP (General System of Preferences) in the GATT grant preferential trade terms (tariff benefits) where there is evidence of positive environmental practice (Dent, 1997; Vogel, 1997). Vogel (1998) shows how EU positions on the environment–trade issue have sought to defend EU firms against foreign environmental measures as well as shelter EU producers from foreign products. Officially the EU opposes levies and other forms of environmental trade barriers, though the European Parliament has advocated reforms which allow non-tariff trade barriers to protect the environment (Golub, 1998a, p. 19). In practice, Golub concludes, 'the EU position on unilateral efforts to force up foreign environmental standards remains finely balanced, if not openly hypocritical' (1998a, p. 19).

Nevertheless, the involvement of the EU in trade bodies offers an unprecedented opportunity for the EU to press upon other nations the importance of the environmental implications of the process of trade liberalisation. Changes that take place in these fora will, in the end, have the deciding effect on how effective an environmental actor the EU is, or how realistic the leadership ambitions of the EU really are.

EU environmental policy and globalisation

It is apparent from the EU's role in the explicit and implicit worlds of global environmental politics that global economic pressures establish the boundaries within which the institutions of the EU operate when it comes to environmental policy. The global political economy impacts upon the role of the EU in global environmental affairs in a number of ways. Firstly, economic prosperity closely correlates with the willingness of the EU to make funds available for environmental programmes. As Golub notes (1998a, p. 1) 'As in other parts of the world, an increasing sensitivity to global economic competition and budgetary constraints has made European governments wary of any form of regulation which might threaten economic growth, foreign investment, export markets and employment creation'.

Secondly, the threat of economic loss in the global game of globalisation is a powerful weapon in the hands of those seeking to check further moves by the EU in the direction of environmental leadership. As Chapter 5 in this book on climate change shows, competitiveness concerns were at the heart of the debate about the carbon/energy tax and were drawn upon by heavy industries opposed to the tax to support their case. The Molitor report of 'independent experts' dominated by industry representatives and ministers pushed the view that environmental regulations pose a threat to the global competitiveness of the EU, arguing that industry is already over-regulated (Golub, 1998a; Vonkeman, 1996). The report forms part of broader chorus of calls from the ERT (European Roundtable of Industrialists) and others demanding deregulation, the scaling back of costly international environmental commitments and the improvement of foreign compliance with existing rules, on the premise that environmental regulation adds to production costs which translate into loss of competitive advantage.

Thirdly, although competitiveness has, of course, been a 'perennial feature' of EU environmental policy-making (Golub, 1998a, p. 1), it is probable that the global context will shape leadership ambitions to an unprecedented degree in the future. It will mean for instance that the EU will push for Conventions that internationalise standards which are already in place within the EU so as to guard against the possibility of 'free-riders' enjoying a comparative advantage over their more regulated European partners. It will also mean that maximum flexibility will be pursued in international conventions. This has been a notable feature of recent deliberations over climate change. Flexibility can also be understood to imply the increasing use of market instruments to tackle pollution rather than traditional 'command and control' state-led, regulatory, top-down approaches. This will involve the use of tradeable permits (already endorsed by the Kyoto Protocol on climate change), joint implementation and voluntary agreements with industry, for example. It would also seem likely that global competition will further encourage the EU's search for 'win–win' scenarios which increase employment, competitiveness and offer environmental gains.

Fourthly, it is possible that environmental standards may become attractive as a way of protecting EU firms from global competition. As Golub argues (1998a, p. 4) 'Negotiating international environmental agreements allows EU members to level the economic playing field and to undermine the effects of pollution havens.' This would avoid previous 'mistakes' whereby, for example, the adoption of stricter discharging standards for phosphate fertiliser raised domestic production costs so that imports of cheap fertilisers from North Africa soared and EU industry relocated to developing countries (Golub, 1998a). As we saw above, internationalising standards that are already in place within the union is a strategy the EU has consistently pursued in its environmental relations with third parties. The overall effect of global economic pressures may then be what Vogel (1997) refers to as a 'California effect' rather than a 'Delaware effect'; an upgrading of environmental standards as a result of liberal trade policy rather than a downgrading, where the EU exports its higher standards to other countries.

Environmentalists would be concerned that the globalisation process which gives rise to environmental problems by accelerating the use and exchange of natural resources, at the same time undermines many of the traditional tools for combating environmental degradation (standard-setting, the use of taxes and subsidies and protectionism). In this regard, the priority given to commercial and trade policy over environmental concerns (indicated by the cases discussed above), the preference for voluntary self-regulation by industry and the insistence upon competitors imposing taxes on their companies before the EU will do the same, are manifestations of powerful global pressures that all circumscribe the scope of EU decision-making on the environment. The outcome of these simultaneous but conflicting pressures towards de-regulation and the internationalisation of regulation will exercise a considerable effect on the future involvement of the EU in international efforts to protect the environment.

The globalisation perspective helps to account for why some polices have been kept off the agenda, why others have been difficult to achieve, and the overall context in which policy has been made. It also further highlights the need to situate theoretical accounts of the EU's role in global affairs within a global political economy perspective.

Prospects

The Dublin declaration of the European Council of 1990 states that the EU must use its position of moral, economic and political authority more effectively in advancing international efforts to promote sustainable development (Liberatore, 1997). The call for such leadership is not new. Before that in 1987 heads of state at the Dublin summit dedicated themselves to establishing a leadership role for the EU in international environmental issues.

Clearly on occasion, when the circumstances are right, the EU can play a leadership role in global environmental politics. This is the case for biodiversity, hazardous waste and most notably, climate change. And yet leadership is a relative term and the EU's policy on climate change is hardly an optimal model. Leadership is defined by the recalcitrance of other industrialised states. Perhaps the best endorsement of the EU's role in global environmental politics is that if it had not been active in these debates, it is more than likely the outcomes would have been less far-reaching.

It is easy, of course, to exaggerate the degree of change that the EU, as a set of institutions separate from the member states, can initiate. The EU only really has an 'independent' foreign policy role in the formation of policy. The implementation takes place at the national level and many of the most important questions will be settled by government ministers at the Council of Ministers. This is especially so in the area of environmental policy. Agenda 21 provides a particularly interesting example of how the multiple levels of authority interrelate. The EU signed up to Agenda 21 at the Rio conference in 1992 and yet it is local authorities that are expected to draft local action plans to implement it aims in the context of national strategies. On one level, therefore, the EU can only be an effective participant in international environmental regimes, if the member states which make up the Union are willing to enforce the commitments at the national and sub-national level.

In this sense, the question of what it is reasonable to expect from the EU as a player in global environmental affairs will be partly answered by the changing membership profile of the EU. It has among its current membership many of the most environmentally advanced countries in the world in the form of Sweden and Denmark though their record to date is mixed. But the accession of members from the CEECS will bring both opportunities and new demands. The extent to which the EU is able to perform a leadership role in the future will depend upon the nature of the constraints imposed upon it by member states new and old.

How the issue of subsidiarity plays will also be of central importance. 'Repatriation' of responsibilities to the national level will ensure that the EU is straightjacketed in its attempts to lead international negotiations on the environment. At the same time, the expanding authority of the EU in other areas may make it increasingly imperative that it speaks on environmental as well as economic issues. The EU will most likely find its niche in acting on transboundary problems of global importance where, symbolically at least, it is best equipped to act. Though the EU could use its position in the global economy to influence many developments that impinge upon environmental protection, a potent combination of fear of reciprocity and the predominance of commercial and trade over sustainability objectives will probably serve to ensure that these opportunities are not exploited in the way environmentalists would hope.

Leadership on questions of global environmental importance is something that the rest of the world expects of the EU and that the EU to some extent expects of itself. In order to perform this function in global affairs the EU needs to be credible. Brinkhorst (1994, p. 617) argues:

> When it is seen that in its own house the EU is credible in 'eliminating unsustainable patterns of production and consumption' [Principle 8 Rio declaration] resistance from other industrialised countries and from the developing world will start to disappear.

The review of the EU in global environmental affairs provided by this chapter suggests the EU, whilst ahead of many of its competitors, is hardly a model of sustainable development that the rest of the world could afford to follow.

5
The Climate Change Policy of the European Union

Introduction

The climate change issue hit the European political agenda amid a set of circumstances which should have benefited the development of an effective policy strategy. Collier (1996a, p. 2) describes the period around the late 1980s as the 'heyday' of EU environmental policy, with the advent of qualified majority voting, the acceptance of 'sustainable growth' as an objective of the EU and increased emphasis on the need to integrate environmental objectives into all other areas of EU policy. Climate change was also identified as a priority in the fifth Environmental Action Plan. Yet, as Wagner (1997, p. 297) notes, 'Climate change arguably represents one of the most serious challenges facing European energy and environmental policy now and in years to come'.

As the only regional economic integration organisation to sign and ratify the UN Framework Convention on Climate Change (UNFCCC), the European Union is in a unique situation regarding the implementation of its commitments (SEC (96) 451 Final). It also has enormous precedent-forming capacity, so that internal decisions have global repercussions. The European Union has hence been under a great deal of pressure to maintain its leadership role in the international negotiations against a background of failed internal initiatives. The pressure is heightened by reflections on what has been achieved six years after the Rio conference in 1992 (Newell, 1997a).

The risk of human-induced climate change was first addressed at the EU level as a research issue in December 1979, when the Council decided to adopt a multiannual EU research programme in the field of climatology (Liberatore, 1994). In 1986 the Fitzsimmons' report determined that there was sufficient scientific evidence of a serious problem to initiate action in three areas: energy saving, reforestation, and the development of measures to prevent tropical forest destruction (Jachtenfuchs, 1996). Nevertheless, despite the fact that according to Jachtenfuchs this report transformed the greenhouse effect into a 'political problem in the EU which required reaction' (1996, p. 95), for almost ten years the issue was primarily regarded as a scientific question by the Commission until, in 1988, it produced a 'Communication to the Council on the greenhouse effect and the EU' (COM (88) 656 Final). The Communication was essentially a 'stocktaking exercise' (Collier, 1996a, p. 3) whose main function was to summarise the state of climate science and the outcome of the Toronto Conference on the Changing Atmosphere, but made no recommendations for immediate action (Wagner, 1997, p. 311). An ad hoc committee was nonetheless established in 1989 containing ten Directorate-Generals including, most prominently, DGXII (Energy) and DGXXI (Indirect Taxation).

Following a European Council meeting in Dublin in June, in October 1990 the Energy and Environment Council of Ministers agreed to return emissions of CO_2 to their 1990 levels by the year 2000 for the EU as a whole, as part of the EU strategy to limit carbon dioxide emissions and to improve energy efficiency (COM(92) 226 final). The division of labour for achieving the target was not made clear at this stage justifying Wynne's (1993) description of the target as an 'ambiguous supranational concoction'. Proposals for equitable target-sharing and allocating individual targets to member states were unsuccessful. Policies suggested to meet the target, which initially included efficiency standards, speed limits, least-cost planning and measures to promote waste recycling were dropped, and the remaining measures significantly scaled down (Collier, 1996a).

The following are some of the policies and measures that *have* been advanced to date as a way of meeting the target. The 1992 Communication on 'A EU strategy to reduce CO_2 emissions' identified a framework directive on energy efficiency (SAVE), a directive on a combined energy/carbon tax, a decision concerning specific actions for greater penetration of renewable energy resources (ALTENER) and a decision concerning a mechanism for monitoring of EU CO_2 emissions and other greenhouse gases. In combination, the carbon/energy tax and the SAVE programme were expected to achieve the bulk of the emissions reductions.

The Carbon Tax Fiasco

Pressure for the adoption of a tax at EU level came from the 'green troika' (Netherlands, Denmark and Germany) who, by threatening to impose unilateral taxation measures, forced a debate on the implementation of a Europe-wide tax to ensure tax harmonisation in line with the single market (see Chapter 1 for more on the troika). The potential effect of the unilateral taxes on national industries drove these states to push for a regional tax covering all member states (Zito, 1995). The carbon/energy tax was also regarded by former Environment Commissioner Carlo Ripa di Meana as the centrepiece of the EU's climate policy strategy. He sought to use the Italian Presidency's favourable attitude towards the tax as a window of opportunity to push the proposal. It was regarded as a way of bringing about a division of burdens within the EU based on countries' use of fossil fuels, whilst bypassing some of the difficulty of apportioning different national targets for each member state. It was intended to encourage the internalisation of environmental externalities in energy prices, and would encourage the further competitive development of renewable sources which is (rhetorically, at least) a further goal of EU energy policy. It also represented a 'foreign policy bid' to demonstrate EU leadership in global environmental affairs (Zito, 1995, p. 432). That the main force behind the tax derived from a desire to raise the international profile of the EU was a source of weakness, because the goal was so diffuse that no particular interests were willing to back the proposal (Zito, 1995).

The combined carbon/energy tax was to be introduced at a level of $3 per barrel of oil equivalent (boe) in 1993, rising by $1/boe annually to a level of $10/boe by the year 2000. The expected reduction in CO_2 emissions to be achieved by the tax was between 3 and 5.5 per cent (COM (92) 226 Final). The early failure of the EU to agree upon this initiative prior to Rio led the Environment Commissioner to declare that he was considering not attending the Rio conference in protest at the lack of political will shown by member states. Some of the fiercest lobbying ever against an EU proposal, saw the tax proposal end up as a watered-down recommendation for national governments to decide whether or not to implement. Industrial groups used a range of arguments centred around the potential loss of competitiveness that the unilateral imposition of the tax might imply, as well as a threat of relocation outside the EU should the policy be pursued further. As a result, exemptions were offered to energy-intensive industries. The exemptions for the six energy-intensive industrial sectors were intended to prevent them vetoing member state support for the proposal at the national level and to pacify the industry-oriented DGs XXI and III. Member states were authorised to grant tax reductions up to 75 per cent to firms whose energy costs amount to at least 8 per cent of the value-added of their products and whose competitiveness might be threatened by the tax. Member states would also have been allowed to grant temporary total exemptions to firms that embarked on substantial efforts to save energy or to reduce CO_2 emissions. This meant, as Collier notes, that 'the largest consumers of energy in the EU would have paid the lowest rates of tax' (Collier 1996a, p. 7). The tax was also made conditional upon the implementation of similar measures by major competitors such as Japan and the United States. A further compromise led to the tax being reformulated so as to apply half to energy and half to carbon content and to electricity output rather than input fuels. The 1994 decision of the Council in Essen does no more than to enable member states to apply a carbon/energy tax 'if [they] so desire' (Collier, 1996b, p. 128). Taken together, these changes weaken the potential impact of the tax such that it is now regarded by most as 'dead in the water'.

The fate of the carbon tax in Europe highlights one of the major constraints on climate change action within the European Union; the political influence of some of the most powerful and economically important industry sectors on the global economy (Newell and Paterson, 1998). That so many sectors will be affected by action to address the problem of global warming brings together in formidable alliance coal, oil, motor, road, petrochemical and heavy industry lobbies, all major employers in the EU, and with the financial backing and channels of access to effectively press their case within the EU (Collie, 1993; Grant, 1993b; Skjærseth, 1994).

The tax also suffered from unfortunate timing, coinciding with a global recession which heightened the sensitivity of Commissioners to the concerns being voiced by the lobbies. There was also concern that the OPEC states might react, or retaliate to such a proposal given the impact of a carbon/energy tax on oil-exporting countries (Vellinga and Grubb, 1993, p. 13). Member states each had their own particular reasons for opposing the introduction of a tax. France was opposed to an energy tax but supported a carbon tax on because it would be relatively unaffected by such a proposal with its extensive nuclear energy programme. The UK, for ideological reasons, was resistant to the notion of extending the EU's taxation powers, and besides Italy, the poorer member states of the EU have opposed the introduction of the tax. Germany had insisted that other member states must make a prior commitment to move to a mandatory, harmonised CO_2/Energy Tax after a four-year voluntary period; a proposal which has been refused both by the UK and France. The fact that the non-mandatory proposal does not contain the conditionality clause whereby similar measures have to be introduced by trading competitors has ensured the continuing opposition of industry federations (Environment Watch Western Europe, 1995b, p. 13). At a meeting of national negotiators in mid-November 1995, Germany dug its heels in on this issue and said that it would rather have no agreement at all than one that did not guarantee the move to a mandatory tax in 2000. Some see the insistence by the Bonn government that energy/CO_2 taxes must be harmonised by 2000 before a voluntary framework can be put in place, as a wrecking tactic on the part of Germany's finance Ministry, who knew this strategy would stall agreement (Environment Watch: Western Europe 1995, p. 10).

The problem is that the carbon/energy tax takes on extra significance when it is recalled that many other non-fiscal options discussed below 'will in fact only attain their full emission abatement potential in a context of higher final energy prices' (SEC

(95) 288 Final). In other words, the impact of other strategies depends upon a tax system being in place which discourages the use of fossil fuels. In addition to this, several member states have indicated that they will not be able to meet their national targets if a EU-wide carbon/energy tax is not put in place. This is the case, for example, with Belgium, Denmark, Germany, Italy and the Netherlands (European Commission, 1996e). Ritt Bjerregaard had stated publicly that the EU cannot avoid a carbon/energy tax if it intends to reduce CO_2 emissions after the year 2000 because energy prices are currently too low to stimulate improvements in efficiency (Environment Watch: Western Europe, 1995d, p. 9). The Commission has reiterated that it has no intention of withdrawing its proposal for a Directive on the CO_2/energy tax, but short-term acceptance is not a realistic prospect. A meeting of EU finance ministers on 11 March 1996 was unable to agree even on guidelines for the Commission to follow in drafting a new initiative. A lot rests on the eventual conclusion of the carbon tax debate. Proposals will probably have to be continually amended until the agreement of the Council can be secured, by which time the proposal may be so weak as be ineffectual in reducing emissions.

However, taxation as a means by which to achieve greenhouse gas reduction targets has not been abandoned. For example, there have been moves to impose an excise tax on energy products. The Commission has proposed a Council Directive restructuring the EU framework for the taxation of energy products (COM (97) 30) which enlarges the scope of the EU minimum-rate system beyond mineral oils to cover all energy products. The 'Monti directive' emanates from DGXXI (indirect taxation), was seconded by DGXI, and aims at harmonising excise duties on energy producers across the EU by setting minimal levels for petrol, gas, oil, kerosene, natural gas, solid energy products and electricity. A number of countries have raised objections to the proposal and the tax, if it manages to overcome member state resistance in ECOFIN (particularly from the UK and southern members), would have little impact on emissions of CO_2, because of the exemptions it carries and the low rate at which it would be set (Environment Watch Western Europe, 1997, p. 9). Juncker, Prime Minister and Finance Minister of Luxembourg, has suggested that the products tax be made optional for member states. As CNE (1998, p. 19) note, 'By working within the existing taxation system, setting initially low minimum levels and emphasising the importance of the proposal in terms of market liberalisation and employment generation, the measure aims to avoid the backlash created by the furore over the carbon tax'. Despite the efforts of the German Presidency to push the directive forward, progress on this issue is likely to be slow as the stalemate at the Council of environmental ministers in March 1999 illustrated.

One of the key obstacles, in addition to those already identified, is the institutional procedure on taxation measures, which means that tax decisions in the Council have to have unanimous approval from all member states. The stalling of even one country therefore can scupper the further development of a proposal. It was this procedure which allowed the UK to stall the carbon tax proposal. Added to this is the fact that Finance Ministers will make the final decision, 'placing the central onus upon institutions that generally have little interest in, or concern about, the climate issue' (Bergesen et al., 1994, p. 27). Furthermore, Article 130s on qualified majority voting for environmental policy explicitly excludes measures 'significantly affecting a member state's choice between different energy sources and the general structure of its energy supply'.

Energy Efficiency and Renewables

The EU has also developed a SAVE (Specific Actions for Vigorous Energy Efficiency) programme which is designed to improve energy efficiency within the EU, in order to reduce CO_2 emissions and improve security of supply. The first SAVE programme, proposed in 1992, set itself a target to improve energy efficiency by 20 per cent between 1985 and 1995. The original SAVE proposal contained plans for legislation to boost energy efficiency efforts across a range of sectors from power generation to buildings, vehicles and household appliances. But by the time it had been adopted in 1993, all the legislative proposals had either been watered down or removed entirely (ENDS, 1996a).

Of 13 separate proposed new laws, only four ever saw the light of day 'owing largely to heavy filibustering by manufacturers of fuel-guzzling appliances coupled with a wave of scepticism about the appropriateness of Brussels becoming involved with non-traded goods items, such as buildings' (Warren, 1996, p. 8). Warren also attributes the demise of the original SAVE proposal to horse-trading in the run-up to the Rio Conference over whether the 'big idea' of the carbon tax should prevail over energy-saving measures. SAVE, he notes, 'was duly sold down the river' (Warren, 1993, p. 57). Energy efficiency standards, in particular, were vulnerable to industry opposition and the lack of legal basis in the Treaty entitling the Commission to propose energy-saving legislation. The EU got less than half-way to the 20 per cent energy efficiency target by 1995, attributed to the collapse in energy prices and a lack of political will among member states (ibid.). Together with Directives on energy efficiency standards, which have for the most part not been agreed, the SAVE programme was intended to reduce EU CO_2 emissions by up to 3 per cent by the year 2000. The SAVE programme was crippled by member states' amendments in relation to subsidiarity, which left most measures to the discretion of individual member states, including the timing and levels at which standards were set. The directive, submitted by a British Presidency keen to assert member state dominance, is broadly stated so that 'Programmes *can* include laws, regulations, economic and administrative instruments, information, education and voluntary agreements' (European Commission, 1993, my emphasis) allowing maximum freedom of manoeuvre for member states. Hence SAVE was turned into a framework directive merely laying out general principles for action to guide member states own programmes and measures, short on targets, deadlines and content. The only proviso was that programmes had to have an impact that could be 'objectively assessed' (Collier, 1996a, p. 9) and that member states report to the Commission every two years on the results of the measures taken.

Yet the few gains achieved by the first SAVE programme were based on the legislative component of the programme which is missing from the SAVE II framework. Instead SAVE II, the second five-year programme to promote energy efficiency, intends to rely on voluntary agreements with equipment manufacturers on labelling and energy standards. Whereas there is no basis in the

Treaty for legislation on energy issues, focus on traded goods allows the SAVE II programme to proceed in an area where there is an established competence (Fee, 1995). The SAVE II proposal includes only a modest target to improve the overall efficiency of energy use in the EU by one per cent over the next five years, and has been subject to significant budget cuts which undermine its ability to meet its self-declared goals. At the May 1996 meeting of European energy ministers, it was agreed that expenditure on the programme should be slashed by over two-thirds. The UK and Germany, despite positive rhetoric on the need to get tough on climate change, were among those states leading the assault on the programme's finances. The SAVE II programme was adopted in December 1996 with an indicative budget of 45 million ECU for five years aimed at improving energy intensity of final consumption by a further one percentage point per annum over that which would otherwise have been attained. For Collier (1997b), the cuts mean that SAVE II is unlikely to be any more effective than SAVE I.

Ironically, there will be more pressure on the SAVE II programme to succeed in the light of member states failure to adopt the proposed carbon tax and the Commission now describes it as the 'corner-stone of the EU's CO_2 reduction strategy' (quoted in Warren, 1996, p. 8; Fee, 1995). A Commission press release describes the programme as making a 'major contribution towards the EU's objective of reducing CO_2 emissions' (Presse, 1995). Given the experience of the first SAVE, where important measures were dropped and detailed requirements removed, this seems unlikely. Furthermore, the degree of flexibility in the programme makes assessing the impact of the SAVE programme very difficult (Collier, 1996a, p. 9).

Standard-setting for appliances in the domestic tertiary sector is also important given that they are responsible for 17 per cent of CO_2 emissions in the EU (SEC (95) 288 Final). The Directive on the energy efficiency of fridges has not proceeded very far, despite the fact that domestic refrigeration appliances are the most important energy-consuming household appliance, with the greatest energy-saving potential. The Directive (as part of the SAVE programme) proposes common standards of electricity consumption in domestic fridges and freezers. The original draft of the Directive was watered down to win the support of EU Industry Commissioner Martin Bangemann. By way of compromise, it was agreed that the second phase of the Directive should only come into effect in 2002 or 2003, four or five years after the first (European Environment, 1995). The target accepted by EU energy ministers was weaker than both the Commission and Parliament had proposed, in part because of intense industry lobbying and a Spanish amendment allowing for more lenient targets for appliances made in hotter countries. The Council Directive of July 1996 requires energy efficiency improvements of 15 per cent over three years and sets minimal energy efficiency standards for refrigerators which come into force in 1999 (ENDs No. 269, 1997). Nevertheless, in general Commission efforts to promote minimum energy efficiency standards have been watered down or delayed due to industry opposition (Wagner, 1997, p. 328).

Mandatory energy efficiency labelling has also been introduced to form part of the Union's efforts to limit CO_2 emissions (Environment Watch: Western Europe, 1995, p. 9). Directives on washing machines and tumble dryers marketed in the EU under Directives dating back to 1992 created the framework for mandatory energy labelling and the provision of product information on household appliances. In May 1997 the Commission adopted a further directive requiring the mandatory labelling of household dishwashers (ENDs, 1997c). If previous experience is any guide, there will be implementation problems, where some parties sell goods without the appropriate labels, despite the labelling being mandatory. Directives are after all binding only in terms of the result to be achieved and not the form or methods undertaken to achieve that goal. In this respect Grubb and Brackley (1991, p. 216) note 'EU proposals on appliance labelling have been applied only haphazardly'.

Given the volume of inter-state transport within the EU, it perhaps makes sense to apply standards at the European level. The diversity in transport systems across Europe is cited as a reason why such a move would be difficult to effect. The size of cars and the range of speed limits differ quite between member states. Transport is also one of the most difficult areas to address given that demand management is a political hot potato. Volumes will increase as part of the pursuit of a single market (see Chapter 7 on air pollution). Nevertheless, transport is an area that must be dealt with. CO_2 emissions from cars currently make up 12 per cent of the EU's total CO_2 emissions (transport more generally makes up 25 per cent of total CO_2 emissions (SEC (95) 288 Final). The draft communication currently circulating on CO_2 and Cars details measures aimed at a 25 per cent reduction by 2008 in the average emissions of new passenger cars compared with the 1995 level. The Commission has also outlined a series of measures (that had already been proposed but not implemented) intended to halve the growth in CO_2 emissions from transport at little or no cost by 2008–12 (European Commission 1998d).

In terms of developing renewable energy within the Union, the Energy Technology Support Programme (THERMIE) programme of the EU sought to strengthen existing measures for promoting the dissemination of better energy conversion and use technologies, and the JOULE (Programme on Joint Opportunities for Unconventional or Long-Term Energies) programme for energy research and development was set up to encourage the development of technologies that would deliver longer-term CO_2 reductions. The JOULE-THERMIE programmes are intended to bring about a 10–20 per cent reduction in CO_2 emissions between 2010 and 2020. Though a recent evaluation of THERMIE indicated that it had made an impact on market shares for energy efficiency technology (European Commission, 1998), clean coal funding under the programme has been criticised by environmental groups for displacing focus on renewables, and for making only a small contribution to lowering CO_2 emissions (Grubb, 1993).

ALTENER (Programme for the Promotion of Renewable Energy Sources) was the proposed Directive to generate support for renewable energy technologies by removing barriers to their successful marketing, by setting (non-binding) goals for the contribution of renewables. The target, however, is a modest one: to increase the contribution of renewables from 4 per cent in 1991 to 8 per cent in 2005 – and the scheme is underfunded. The final decision bringing the ALTENER into being in January 1993 contained specific targets but no substantial tools for implementation. It is also a 'moot point' whether even the existing 1 per cent reduction target in CO_2 emissions will be met (Haigh, 1996; Collier,

1997a). The Energy Council meeting of 11 May 1998 adopted ALTENER II for 1998–1999 with a budget of only 22 million ECU (the Parliament and Commission had proposed more than 30 million). A Commission White Paper published in December 1997 called for a doubling of the proportion of EU energy needs supplied by renewables to 12 per cent by 2010, whereas they currently supply just 5.3 per cent of EU energy consumption. This new target would deliver one-third of the 15 per cent CO_2 reduction target (ENDS, 1997j) and includes a 'campaign for take-off' to help lift the share for market renewables by promoting photovoltaic schemes, biomass projects and wind farms (European Commission 1997c). The UK, France and Germany have opposed any specific commitment to increasing the share of renewables to 12 per cent by 2010, however, and the Council accepts the Commission's target only by way of guidance; making it a voluntary target for member states. Efforts to promote renewables are also undermined by the continued use of subsidies to fossil fuels. A Greenpeace report on 'Energy subsidies in Europe' showed that more than 90 per cent of direct subsidies from European governments to the energy industry go to fossil fuels (Greenpeace, 1997).

In terms of managing demand for energy, the IRP (Integrated Resource Planning) Directive proposal adopted by the Commission has evolved into a toothless instrument. In theory it has the potential to satisfy all three imperatives of EU energy policy: improving economic competitiveness, increasing security of supply and protecting the environment. The Draft Directive (COM (95) 369 Final (European Commission 1995c); to introduce 'rational planning techniques' was put forward under the SAVE programme (see above), intended to ensure that equal consideration is given to demand-side as well as supply-side options in meeting consumer demand. In other words, it calls on electricity and gas companies to examine the economic potential of energy efficiency measures on an equal footing with the traditional option of increasing their supply capacity. Suppliers and distributors would be required to prepare strategic plans indicating how they will meet future demand using alternatives. IRP is seen as an important way of enabling the EU to meet its stabilisation commitment (EC-Energy Inform, 1995) potentially reducing 1994 emissions of CO_2 by 3 per cent (a reduction of between 4 and 8 per cent in the level of energy demand). Indeed the Commission has invoked the stabilisation target as one of the key reasons for implementing the Directive (EC-Energy Inform 1995: 5).

The Directive was given overall support by the European Parliament Energy Committee, though it stated a preference for a non-binding recommendation rather than a Directive to increase its chances of adoption by member states. As things currently stand, 'it appears that ministerial discussions on the measure will remain deadlocked for some time' (*Environment Policy Reporter*, 1996). The proposal has not yet been submitted to the Council and ministers have set no deadline to adopt the measure. This dithering takes place against the background of a warning by UK MEP Eryl McNally, who drafted the Directive, that Rational Planning Techniques are 'the last chance to make energy efficiency happen within the marketplace structure' (Collins, 1996, p. 3), and resistance from utilities' insistence on the grounds that the proposal is incompatible with aspects of the legislation on the internal energy market. The Directive could easily fall prey to a battle of ideology between free marketeers and state interventionists. Its passage will not be easy given the supply-side preoccupation of the Commission and in the face of objections from powerful industry bodies such as EURELECTRIC. Neither will the future structure of a liberalised electricity and gas sector create the right incentives towards demand-side management, a factor which will undermine its overall effectiveness (Collier, 1997b).

The most persuasive argument in favour of Rational Planning Techniques is that other initiatives, such as the carbon/energy tax, efficiency standards, labelling and voluntary agreements, will only be able to achieve the full energy efficiency potential if they are complemented by an initiative addressing energy (especially electricity and gas) suppliers and distribution companies.

Given this background of resistance to regulatory approaches, voluntary agreements are a popular option with energy industries, who have pressed their importance upon governments as an alternative to regulatory mechanisms. There is some evidence, however, that they have not been properly implemented (Vellinga and Grubb, 1993, p. 19). Experience at the national level (such as in the Netherlands and Germany) also suggests the need for a regulatory framework to be in place to guide the initiatives, which is currently lacking within the EU. Nevertheless, manufacturers are keen to press upon policy-makers their effectiveness and have already agreed to fairly substantial cuts in emissions. European car manufacturers for example have committed themselves to reducing CO_2 emissions from new vehicles by 10 per cent between 1993 and the year 2005 as a way of pre-empting stiffer mandatory codes which the EP is calling for. Often however, the agreements fill a vacuum left by the failure of mandatory standards. The voluntary agreements drawn up following negotiations with the Commission by trade associations to reduce the energy consumption of washing machines, videos and TVs provide an example of this (ENDS, 1997i; ENDS, 1997j). For some member states such agreements will form an important part of an overall package of measures. The UK is assuming for instance that a not insignificant share of its 12.5 per cent target will be reached through such agreements.

Commenting on the key tenets of the EU's climate policy, Collier reported in 1996b that the proposal for a carbon/energy tax had been blocked, the SAVE programme had been turned into an ineffective framework directive, the ALTENER programme was under-resourced and based on non-binding targets, and proposals for reducing CO_2 emissions from cars and least-cost planning in the energy sector had been delayed (Collier, 1996a). The establishment of a monitoring mechanism represents the only substantive piece of EU legislation to have been adopted by the Council so far. This requires member states to devise, publish and implement national programmes for limiting CO_2 emissions in order to help achieve the EU's stabilisation target. It also contains provision for the evaluation of these national programmes by the Commission establishing an inventory of greenhouse emissions under the FCCC (Wagner, 1997). The central tenets of EU climate policy have all been weakened to such an extent that their effect upon levels of emissions within the EU will be minimal. As Grubb (1995, p. 6) notes 'the post-Maastricht and post-Rio reality hangover has neutered almost all elements of this strategy'. Even the promotion of energy efficiency and renewable energies, thought to provide 'a politically and economically palatable tool for the EU greenhouse policy' (Liberatore 1994, p. 198) have been problematic. The fact that the EU did not manage to keep to the stabilisation target in a time of recession suggests that capping the growth in emissions during periods of renewed economic growth will be a far greater challenge. The Commission's Second Evaluation of National Communications under the Climate Convention declares, for example, that anticipated economic growth of 3.3 per cent between 1995 and the year 2000 will lead to an increase in emissions over and above the national programmes implemented by member states (European Commission, 1996e).

The result of the lack of progress on the various aspects of the EU's climate policy strategy is that responsibility for meeting the EU's stabilisation target will fall upon member states (Collier, 1997a). Most member states, however, have failed to meet their stabilisation targets with only the UK, Germany (for reasons outlined below) and Luxembourg being successful. This means that the 'EU's role has been relegated to monitoring what its member states choose to do unilaterally, and projecting the consequences' (Grubb 1995, p. 6). Subsidiarity has sharply constrained the evolution of a genuinely EU-level climate policy. The principle has been invoked as a justification for opposition to the carbon tax (in the case of the UK) and is largely responsible for the scaling down of the SAVE proposal into a framework directive (Collier, 1993). Whilst the renationalisation of climate policy is not detrimental to the overall effectiveness of EU climate policy in all cases, it does mean that a comprehensive and coherent policy cannot then be guaranteed in all member states.

While the relationship between national and European-level responsibilities is worked through, this would appear to be the pattern of things; the EU sets goals and offers guidance, while member states select content. In terms of meeting the requirements of the Climate Convention, this is problematic because, as the first evaluation illustrated, only two countries (the Netherlands and Denmark) provided evidence of a full programme with specific measures to achieve their target. Many member states merely submitted plans indicating the technical possibility of meeting the target accompanied with lists of potential policies that might achieve it. There is therefore insufficient information about the implementation of measures for the Commission to evaluate the effectiveness of individual countries policies (Collier and Löfstedt, 1997, p. 12). Most policies listed are those that were already in place, marginally extended, or put in place to achieve other policy goals and re-packaged as climate policies. The vagueness of the information contained in member states' national communications is indicated by the Second Evaluation which explains that the Commission has been 'unable to establish a EU projection for the year 2000 on the basis of information supplied by Member States given the breadth of the methodologies and hypotheses used' (COM (96) 91 Final). The Commission concedes 'So far the monitoring mechanism has been relatively weak and data has been late' (European Commission, 1998g). In effect, this means that the EU is not meeting its commitments under the FCCC because projections have not been adequately harmonised to assess whether or not national policies will be sufficient to fulfil the joint stabilisation target.

Internal politics

One of the principal difficulties in developing an effective climate change strategy in the Union is that it corresponds so closely to the politically sensitive area of energy policy in which the EU does not have an established competence. Member states retain control over decisions made in this area on the whole, and there is a strong reluctance to permit the extension of competence. Energy policy is subject to a number of 'structural interests' (Matláry, 1997, p. 3) and, as such, is one of the weakest areas of EU policy. Environmental issues have become one of the principal vehicles through which the consolidation of EU-wide energy policy has been sought.

The context of the three pillars of EU energy policy, namely security of supply, environmental protection and deregulation (the creation of an internal energy market) shape the contours of EU climate change policy in important ways. Climate policy has hinged on measures in two areas that remain the prerogative of member states, namely energy and fiscal policy (Collier, 1996b) where there is a lack of clear EU competence. This has implications for measures in the area of energy efficiency and renewable energy and, of course, for the carbon/energy tax. The 'continuing power struggle' (Haigh, 1996, p. 155) between the institutions of the EU and member states, as the examples above illustrate, was one of the main reasons behind the decline of Commission initiatives related to climate change. The objectives of the SAVE programme were argued to be best achieved at the national level and subsidiarity was used by the UK government as a basis for opposing the carbon tax on the grounds that it already had equivalent measures in place in the form of VAT on fuel.

The central challenge is to implement policies at many different levels. As Bergesen et al. note, 'There is a tension between sovereignty, the needs of collective responsibility and the harmonisation of responses where appropriate according to the principle of subsidiarity' (Bergesen et al., 1994, p. 6). For while there is evidently a role for the Commission in proposing legislation on efficiency standards for tradeable goods and harmonising fiscal measures, and also in coordinating European wide research programmes, other measures in non-tradeable areas such as building insulation standards are less appropriately addressed at EU level. Appliance standards and energy taxes require EU-level harmonisation to allow for the proper functioning of the internal market (Collier, 1996b). Instruments that impinge upon the single market in areas relating to external trade or agriculture (such as support for biomass), for example, also have to be decided at EU level.

One important role for the Commission is to ensure that national programmes add up to the EU target and to 'sound the alarm' if they do not (Haigh, 1996, p. 166). It is unclear exactly what forms of pressure will be brought to bear upon states whose failure to fulfil their obligations undermines the EU goal, but it should be acknowledged that a number of EU countries would fail to implement policies were it not for EU-level action (Collier, 1996b). As Collier (1997a, p. 125) notes in the case of Italy:

> Given the resistance at national level to the formulation of such measures, the role of the EU is fundamentally important as an institution to give policy direction through binding instruments such as directives, in defining regulations, standards and in proposing carbon/energy taxes.

However, member states will be reluctant to cede more power to the EU to determine policy within the EU, because this simultaneously becomes a competence to agree rules on that subject in international fora (Haigh, 1996).

It has been suggested that if all member states took their obligations under the Climate Convention seriously, EU action could almost be restricted to coordination and funding for research, development and demonstration (Collier, 1996b). This is clear from the Commission communication on the EU's post-Kyoto strategy, which emphasises an EU role in providing a framework for implementation whereby member states provide detailed information to the Commission on how they intend to meet their individual targets and the contribution they expect from the EU. It foresees a role for itself as a 'forum for exchange' of experience and research on climate change and in the event of a tradeable permit system being developed, to oversee the trading system to ensure compatibility with the single market (in terms of avoiding discrimination and distortion of competition). In the absence of effective action and enforcement on the part of member states, however, the role of the EU needs to be more proactive if the goals it has set for itself are to remain realistic.

In the absence of a clearly defined area of exclusive EU competence for climate change, and in the absence of clear obligations detailing specific action, it is extremely difficult, nevertheless, to isolate EU and member state obligations (Macrory and Hession, 1996). As Haigh (1996, p. 184) pertinently notes, 'Until Member States take global warming seriously and substantively seek to mend their own institutional way, the Commission can promise and prepare, but no effective institutional transformation can be expected.' The balance of power on climate policy between the EU and member states will be partially determined by the extent of the use of market mechanisms in abatement efforts. Traditional regulatory approaches (the traditional purview of the EU) will be restricted to activities such as banning the use of HFCs for self-chilling cans or the adoption of efficiency standards for electrical equipment, while flexible market mechanisms for different economic sectors will be largely established at the national level.

The battle over policy authority continues within the EU too.

The European Parliament (EP) has sought to raise its profile in the policy-making process through adopting a strong position on environmental issues. For example, the CERT (the Committee for Energy, Research and Technology) has targeted the transport sector for reform, takes a positive view of the carbon tax and stresses the need for energy efficiency provisions in the European energy charter (Matláry, 1997). The EP is also supportive of a common energy policy with environmental protection as a guiding goal. The enhanced role of the Parliament as a result of Maastricht may have the effect of tipping the overall balance slightly in favour of a proactive position on climate change given the active past record of the environment committee and the representation of the Greens in Parliament (Boehmer-Christiansen, 1995). The Global Legislators Organisation for a Balanced Environment (GLOBE) group within the Parliament have also been very active in trying to push EU climate policy forward, particularly Carlos Pimenta and Tom Spencer (MEP) (GLOBE, 1997). Nevertheless, despite its good intentions, 'the formal role of the EP in the energy policy-making process must be considered fairly insignificant' (Matláry 1997, p. 127). The EP is still a largely consultative body and despite the co-decision procedure that endows it with the power to amend or even veto a proposal, it has no direct influence over the final outcome and therefore continues to play a secondary role to the Council (Wagner, 1997).

Moreover, despite the fact the Treaty on European Union (Maastricht Treaty) introduced qualified majority voting for environmental measures, it also confirms that unanimous voting continues to operate in policy areas of a fiscal nature and measures affecting a member state's choice between different energy sources and the general structure of its energy supply. Climate policy-making will therefore continue to be dogged by slower, more incremental consensus-building decision-making processes subject to the veto power of objectors.

Interest groups also play a key role in EU climate change policy, as the case of the carbon tax above makes abundantly clear. Interest groups are invited to have input at the earliest stages of policy and are represented on many of the committees through which draft proposals pass. Environmental NGOs have not generally been a prominent force in these discussions (Matláry, 1997) with the possible exception of specialised umbrella groups like Climate Network Europe (Newell and Grant, forthcoming). The big industry players are 'Euro-federations' such as Europia (oil) Eurogas (gas) and Eurelectric (electricity) and UNICE (representing a range of heavy industries as an employer and employee organisation) who provide technical expertise and use their degree of market access to press upon Commissioners their preferred policy options (see Chapter 2). Industry groups also have closer relations with the more influential DGs in energy policy issues. DGXII, the energy Directorate, is said to be very responsive to energy industry interests (Matláry, 1997, p. 107).

The problem of coordinating an effective response to climate change within the EU is further exacerbated by the range of DGs involved in the formulation of policy. The Commission interservice group on the carbon tax in the run up to Rio included DGXXI (Tax), DGIV (Competition), DGXVII (Energy), DGXVI (Regional Policy), DGXII (Science and Research) and DGXV (Financial Institutions), as well as DG XI for Environment (Bretherton and Vogler, 1997). The ad hoc committee established in the Commission in 1989 to address the climate change issue included ten DGs, of which environment, energy and indirect taxation became the most prominent (Collier, 1996). The involvement of the tax directorate had a negative impact upon the fate of the carbon tax given, as Zito notes, 'DGXXI had been one of the least enthusiastic about the tax and its head Christine Scrivener, had publicly questioned the merits of the proposal' (Zito, 1995, p. 441).

The relationships described by these interactions can be thought of in terms of a three- or four-level game played simultaneously between the member states and the EU, among the institutions of the EU in their internal power struggle, between the different directorates and possibly also between the various directorates and interest groups (Matláry, 1997). Each actor seeks to claim for itself a role in the formulation and implementation of climate policy so that a multi-level bureaucratic turf war is the inevitable result (see also Chapter 4).

The disparities in levels of development between the member states of EU (particularly between North and South) create extra challenges for a cohesive climate policy. Individual member states differ in terms of the domestic pressures they are under, their fuel mixes and regulatory practices, views towards Europe and energy policy histories (Collier, 1997). 'Leaders' and 'laggards' have emerged on the issue reflecting a broadly North–South cleavage, so that while countries like Spain and Italy seem to follow the path laid down by Brussels, Germany and the UK appear more proactive in establishing the terms of debate (Collier, 1997a, p. 163).

The EU seeks to reconcile these differences through burden-sharing, whereby international commitments are implemented collectively by distributing different responsibilities amongst member states. This provides poorer members of the EU such as Greece, Ireland, Portugal and Spain (the 'cohesion' countries) 'ecological space' in which to increase their emissions, whilst the more industrialised member states are expected to accommodate this emissions growth. The Council agreed in March 1997 that some member states should reduce their emissions, others should stabilise them and five member states would be allowed to increase their emissions.

It is important to note, however, that burden-sharing only works if countries are willing to accept burdens. Most member states that will be expected to shoulder accept burdens have not shown willing to date. The likely failure of the EU to meet its stabilisation target is especially embarrassing for the less developed member states who signed the Climate Convention with the expectation that their emissions growth would be accommodated under the 'umbrella' EU target. Indeed Spain insisted on the EU signing the FCCC as a bloc in order to allow it to increase its emissions (Villot, 1997). Article 4 of the Kyoto Protocol makes clear that no member state will be in breach of the Protocol providing the EU as a whole is in compliance. In the event of failure to reach the total combined level of emissions, each party will be responsible for its own level of emissions set out in an agreement communicated to the Climate Convention secretariat (European Commission, 1998b).

The Kyoto Protocol creates the added problem of deciding what mix of policies member states within the EU 'bubble' can adopt to meet the target and whether a ceiling should be set restricting the use of Kyoto mechanisms. This latter issue has fragmented the usual alliances on the climate issue with Sweden, Finland and Netherlands siding with Ireland in pushing for greater flexibility and Austria, Denmark and Germany seeking to impose a 50 per cent ceiling on the use of the Kyoto mechanisms (*ENDS Daily*, 24 March 1999).

National policy responses

Momentum for a change in the direction in EU climate policy may come from particular member states, though the source of potential leadership is not easily determined. The UK continues to play an obstructionist role in discussions both on the carbon/energy tax and energy efficiency standards in appliances and has failed to implement measures such as full-rate VAT on domestic fuel and the Energy Saving Trust's work programme (ENDs, 1997a). The anticipated change of direction under the more pro-European 'New Labour' government has not been dramatic. Prime Minister Tony Blair has said that his promise to reduce the UK's emissions of carbon dioxide alone by 20 per cent by 2010 from 1990 levels is not a conditional target (ENDs No. 269, 1997). At the June Environment Council meeting in 1998, however, Britain was willing to adopt a commitment of only a 12.5 per cent reduction in CO_2 emissions.

Germany, as was noted above, has used its Presidency to inject new life into taxation initiatives and to try and gain agreement about the degree to which the Kyoto mechanisms should play a role in delivering the EU's reduction commitments. It remains, however, the largest emitter of CO_2 in the EU, and the leadership mileage that was gained from the 'wall-fall effect' of benefiting from former East Germany's de-industrialisation and hence temporary de-carbonisation, is likely to exhaust in the near future as transport and industry emissions threaten to overwhelm the government's ability to meet its target (Bach, 1995). Germany's schizophrenic behaviour in pushing to axe the budget for the SAVE II programme at the Council of Ministers whilst at 'home' pushing for one of the most advanced greenhouse gas abatement policies in the EU, has not gone unnoticed. The problem, however, is that the ability of the EU as a whole to meet the stabilisation objective rests on Germany keeping to the anticipated

12 per cent reduction in national emissions by the year 2000 as way of counterbalancing expected increases in other member states (Heller, 1998).

France is also finding it increasing difficult to hide behind the defence that its nuclear energy programme makes it a leader in the greenhouse debate. Oil consumption, largely from the transport sector, is increasing, and France is now ranked eleventh in the world table of energy consumers.

Many national-level initiatives have also come into conflict with EU competition policy. Where innovative measures have been developed, pressure has been brought to bear in ensuring their defeat. The German 'feed-in law', which allows surplus electricity generated by renewable energy forms to be fed back into the national grid at favourable rates, has been challenged under competition policy as favouring one sector over another. Similarly, Denmark was advised by the Commission that its proposal for a national CO_2 tax would contravene the rules of the EU's harmonised system of excises on mineral oils. The few initiatives that have been taken, therefore, have been frustrated by the wider market integration priorities of the EU.

Integration with other policy sectors

The lack of coordination across different sectors of the EU has also weakened the impact of climate policy measures. Attempts by the environment Directorate to initiate policies, for example, on climate change are often offset by the growth in emissions from the transport sector. The opening sentence of an article on EU roads policy captures this incoherence well: 'The European Parliament has voted to introduce tougher environmental safeguards into the EU's plans for trans-European transport networks, but also to include more roads in its programme' (Environment Watch: Western Europe, 1995c, p. 13).

The EU's energy market liberalisation plans, as the case of the UK illustrates, will also probably have the effect of lowering the price of fuels and therefore encouraging greater consumption of fossil fuels. The White Paper on Energy Policy issued by the European Commission in December 1995 illustrates this dilemma (European Commission, 1995f). The paper, whilst acknowledging that without major interventions in the energy market, EU CO_2 emissions will increase substantially over the next two decades, fails to reconcile a basic conflict between the drive to liberalise energy markets and therefore push down prices and the 'weakening momentum towards energy conservation' (ENDS Report, 1996). The argument of Commission officials is that environmental concerns should not interfere with the removal of inefficiency in the energy sector which is perceived as a good thing per se (Collier, 1993a). The lack of consideration given to the implications of the internal energy market (IEM) proposals for the greenhouse gas emissions of the EU is also a function of the fact that the IEM proposals date back to the mid-1980s when global warming was not established as an issue on the policy agenda (ibid.).

The European Energy Charter, which aims to stabilise investment rules for energy exploration, production and investment in Europe, could exacerbate this trend, where 'low energy prices brought about by the Charter could counteract all efforts to save energy and promote renewable energies' (Jachtenfuchs, 1996, p. 132). The Charter will ultimately carry protocols on environment and energy, towards the charter's aim to 'promote an efficient energy market with due attention being paid to the environment' (Matláry, 1997, p. 54). Collier (1997a, p. 50) concludes, however, 'The whole IEM discussion has proceeded without much reference to the climate change issue.'

The process of further economic integration of which infrastructural development and market liberalisation are a part, will also have repercussions for the effectiveness of climate policy in terms of increases in production and transport that will accompany the further deepening of the internal market. Trans-frontier lorry traffic is likely to increase by between 30 and 50 per cent as a result of the removal of customs and border controls (Weale and Williams, 1992, p. 52). Climate change perhaps highlights more clearly than other areas the discrepancy between the economic and environmental objectives of the EU. Collier notes (1996a, p. 3) 'it . . . has to be acknowledged that there may be some inherent incompatibilities between the pursuit of unfettered economic growth and the aim to reduce greenhouse gas emissions'.

The broader point is that political decisions across a range of sectors which touch upon climate change are proceeding largely uninformed by the need to integrate climate concerns in all areas of policy. As Weale and Williams (1992, p. 49) note, 'despite the formal stress placed upon the integration of environmental concerns into the making of a wide range of EU policies, it can be argued that the implementation of integration has been a faltering and haphazard affair, without serious resonance in the central policy activities of the EC'. Because of its scope and the multiplicity of its sources, climate change interfaces with numerous other aspects of environmental and energy policy, so that the breadth of de facto climate policy is considerable even if the EU does not recognise it as such.

Part of this failure to ensure that activities undertaken elsewhere in the European policy-making process are sensitive to the goals of climate protection relates to the bureaucratic weakness of DGXI and its limited administrative resources in ensuring the proper integration of objectives and guarding against the sectorisation of environmental policy (Weale and Williams, 1992; Dent, 1997). Because of its weak overall position in the formulation of policy, DGXI is forced to forge alliances with other DGs in the Commission. The alliance between Ripa di Meana and Antonio Cardoso e Cunha, the energy commissioner, to support the carbon tax provides just one example. This alliance acted as an 'entrepreneurial coalition' pushing the tax proposal though the Commission (Zito, 1995).

Explanations

Climate policy is clearly guided by wider objectives of EU policy, such as the need to tackle unemployment. Indeed it may the case that if energy taxes or other taxes on consumption are to be introduced it will be in order to reduce taxes on labour as a way of reducing unemployment rather than to reduce the output of greenhouse gases per se. Heller (1998) shows how the Commission attempted to package the carbon tax as offering a 'double dividend' of increased growth and employment by switching taxes away from firm's use of labour towards their use of resources. Wilkinson also argues (1997, p. 164) 'energy sector attempts to reduce energy demand simultaneously serve a number of different policy objectives, the classic "win–win" situation and are therefore more readily acceptable to DG'. Repeated reference to 'no-regrets' options in Commission documents are a further manifestation of this trend, where action is taken which can be justified on grounds other than the need to address climate change alone (such as competitiveness, security of supply, etc.). The greenhouse effect has also provided a useful justification for reviving older policies for achieving energy efficiency and the other goals of energy policy which had been considered unsuccessful to date (Jachtenfuchs, 1996, p. 98). Energy-saving measures in particular seemed to hold out the possibility of relaunching EU energy policy.

Climate policy will continue to be interpreted within the terms of existing policy 'frames' and institutional needs and will succeed to the extent to which it accords with the objectives and practices of those established ways of interpreting 'new' policy problems (Jachtenfuchs and Huber, 1993; Jachtenfuchs, 1996). This accounts for the early treatment of climate change as a scientific research problem and latter incorporation within the logic of consolidating and extending a common energy policy (ibid.). General pressures towards integration and the use of economic instruments for abating pollution (rather than traditional 'command and control' instruments) also strongly informed the interpretative frame within which climate change came to be understood. The frame acts as a boundary, validating certain policy options and marginalising others in a incremental process of 'filtering'. It helps to explain why policy measures which address fundamental patterns of resource intensive energy consumption and production will not be acceptable because they challenge core activities within the EU. What is notable in the climate case is that there were multiple competing frames 'sponsored' by different directorates-general each seeking to assert their preferred way of addressing the climate change problem. DGXII rejected any notion of reducing the consumption of energy (supply frame) and the use of taxes to achieve to such a goal, while DGII (Economics and Finance) were keen to see economic instruments applied. For DGXI, the leadership frame took priority and taking action on climate change was important for its position within the Commission and for the EU's position in the international negotiations on the subject.

The abstention of both the US and Japan from leadership ambitions created a window of opportunity for the EU to project itself as a serious international player in environmental affairs.

Policy outcomes reflect attempts to reconcile these different approaches under a new frame, though the bureaucratic powerbase of each DG clearly also affects its ability to successfully project its preferred understanding. The carbon tax debate, for example, shows how the competitiveness frame sponsored by Bangemann (internal market) and Scrivener (taxation) won out over other concerns in the run-up to the Rio conference, ensuring the tax was made conditional upon the adoption of similar measures by the EU's major competitors. The success of the competitiveness frame also has to viewed in the light of the trend towards greater reliance upon economic rather than scientific input and policy advice as a proposal proceeds through the policy cycle (Liberatore, 1994). Hence the Council relied much more upon cost-benefit estimations than climate modelling and risk assessments (Jachtenfuchs and Huber, 1993). The political influence of each DG also changes as the policy under consideration alters. When the main focus of climate change policy shifted from energy efficiency to taxation, DGXII moved towards the periphery of policy development, while DGII (Economic Analysis) and DGXXI (Indirect Taxation) became more important because of its expertise in issues pertinent to that policy option (Huber, 1997).

Politically uncontroversial frames such as 'no-regrets' nevertheless serve the purpose of providing an umbrella concept that competing DGs can subscribe to; they form part of a consensus-building process that permits policy to develop. Similarly, the demand for environmental leadership and the continued efforts within some parts of the EU to call for a carbon/energy tax are couched in terms of their ability to promote the wider integration objectives of the EU to secure the widest possible basis of support (Jachtenfuchs and Huber, 1993). Problem definitions need to be operationalisable, manageable and unthreatening to existing core practices and goals in order not to undermine the legitimacy of the organisation addressing them:

> The primacy of the integration frame means that if decisions are to affect the institutional balance of the EC, the first cri-

> terion for adopting, modifying or rejecting them is not whether or not they contribute to solving the immediate problem or whether they are too expensive or too bureaucratic, but only whether the change in the institutional balance they are expected to entail is acceptable to the participants or not. (Ibid., p. 54)

Prospects

As far as potential leadership from DGXI goes, Danish former Environment Commissioner Bjerregaard had shown a lack of leadership on this issue and was thought to be very marginalised in the overall policy process. Bjerregaard had an inexperienced team of advisors, and lost many of the internal battles over environmental legislation. Her frequent defeats earned her the tag 'industry's best friend'. The isolation of her cabinet in the overall policy process was compounded by former Commission President Jacques Santer's professed lack of interest in environmental issues (Environment-Watch Western Europe, 1996, p. 1).

The accession of new member states which have fairly positive environmental track records to the EU, such as Sweden, may hold out the prospect of a strengthened climate strategy, but the limits of what can be achieved by one or two member states are very real. Alliances with states such as other new member Austria may form the basis of a 'green' group within the EU which would be sufficiently assertive to adopt a leadership role on climate policy. The newer members are more supportive of the use of fiscal instrument to combat environmental problems, suggesting promise in that area (Bergesen et al., 1994; Wagner, 1997). Yet it should be recalled that other new members such as Finland have only agreed to stall the increase in energy-related CO_2 emissions by the end of the 1990s and Austria, Finland and Sweden all have low per capita emissions, making further reductions a difficult exercise. It is also likely that with the accession of new members from Eastern and Central Europe (given the surplus of carbon-heavy brown coal in those areas) there is a possibility of veto partners for other member states seeking to delay taking action on climate change, and hence the net effect may be to reduce the likelihood of a deeper strategy. However, others (for example,

Heller, 1997) view the accession of member states from the East as an opportunity to be exploited where low cost investments with higher emissions savings are to be found in abundance.

The rotating Presidency of the Commission may also provide the necessary impetus to guide the EU towards a more effective climate change strategy. The Irish Presidency did not make much of an impact, but the Dutch, with a comparably more progressive national position on climate change, made the issue a top priority. The Dutch were successful in agreeing the 15 per cent cut in emissions by the year 2010 amongst member states for the basket of greenhouse gases (carbon dioxide, nitrous oxide, and methane) disagregated by country and hence at least partially helped to address the burden-sharing issue (see above). The impact that a more progressive Presidency can have upon the overall shape of climate strategy should also not be overestimated, as incoming Presidencies are presented with a backlog of unfinished business which they are expected to complete (Wurzel, 1996). Germany, for example, has inherited from Austria the unenviable task of pushing forward the energy products tax during their Presidency.

As things currently stand, the EU does not have a climate policy per se. There is not a single piece of legislation in the form of a Directive which has as its *sole* aim the combating of climate change. As was mentioned above, policies that were conceived as being part of an EU climate policy strategy have largely been vetoed or heavily gutted and alternatives are scarce. The role of the EU has been reduced to that of overseeing that member states meet their obligations under the Climate Convention and coordinating a common response. As Heller (1998, p. 121) notes, the failure of EU-wide programmes 'has returned the onus of policy to member states unprepared or unwilling thus far to bear it'. Given that many of the 'no-regrets' options have not yet been developed, it is difficult at this stage to foresee the EU putting in place a series of initiatives designed exclusively to combat climate change. The Commission suggests that there is cost-effective technical potential for emissions reductions of up to 20 per cent by the year 2010 for the EU as a whole (European Commission, 1995b). Yet as it also notes, 'this technical potential will only be achieved if there is a political will to adopt a wide-ranging package of measures, effective enough to remove all existing barriers to CO_2-limiting investments' (ibid.). It is the political issues touched upon in this chapter rather than scientific and technical constraints that will shape future EU climate policy.

Perhaps the harshest indictment of EU climate policy is Bergesen et al.'s observation (1994, p. 27) that 'the Union and its member states have actually impeded each other's efforts to address climate change: that so far, the sum of responses may be less than the parts might have individually achieved'. Member states called on the Commission to develop proposals for meeting the CO_2 target and then blocked or neutered virtually all substantive Commission initiatives on climate policy. Grubb and Brackley (1991) note though, that while the EU may retard the leaders, it will act as a spur for most countries. EU support for renewable energy development is also stronger than that which exists in most national governments for example. Hence, in some areas, 'The response of the whole may well be greater than the sum of the parts' (ibid., p. 230). It should be recognised, therefore, that a number of EU countries would fail to implement policies either nationally or locally were it not for EU-level action (Collier, 1996a). Policy will continue to be driven by those member states (Germany, Austria, Denmark, Netherlands) that have set themselves individual targets that go beyond the EU target. The analysis above, however, suggests that the lessons in the development of effective climate policy that one would want to replicate across the Union based on the experiences of these countries are limited. The story across member states is one of failed tax initiatives as a result of industry lobbying and government's acceptance of voluntary agreements as a basis for exemption from taxes, cuts in energy prices condemning attempts at demand-side management, and a general failure to address rising emissions from the transport sector. Trends towards deregulation and privatisation have meant that governments have lost a lot of control over energy companies so that (Collier, 1997b, p. 104) 'regulators have not been given a strong environmental remit and certain issues, such as energy efficiency, clearly fall between regulatory chairs'.

'Effectively, the EU climate change strategy consists of some guidelines on energy efficiency measures, targets for renewable energies and support programmes for the development of energy technologies. None of these are expected to make a substantial impact on emissions' (Collier and Löfstedt, 1997, p. 10). Reducing emissions of greenhouse gases remains the most intractable problem facing the EU today. The climate change issue has failed in most respects therefore to command a policy status of its own, but has been defined in relation to pre-existing policy needs and institutional practices. There has been little attempt to reflect on the unique challenge that it poses or the extent to which it challenges current goals and priorities. Emission reductions that have occurred have come about as an incidental by-product of economic restructuring.

The course that future policy takes will also be strongly influenced by wider international economic and political developments. As the Commission Communication on Kyoto makes clear (European Commission, 1997b), 'The underlying assumption for the EU policy on climate change is that other industrialised countries undertake comparable commitments to cut greenhouse gas emissions'. Not only will wider international events affect the EU, but the climate policies of the EU have significant repercussions for how the rest of the world responds to the issue. As Michael Grubb argues 'given the ingrained recalcitrance of the United States and to some extent Japan, the EU is at present still the only credible candidate to lead future efforts' (Grubb, 1995, p. 5). Yet, unless economically more advanced and politically cohesive regions of the world like Western Europe show that action on climate change is economically feasible and ecologically imperative, it is both unrealistic and unreasonable to expect that other less well-developed regions will develop similar strategies. As the Commission notes, failure to reach the stabilisation objective by 2000 'could damage our ability to convince developing nations in particular to pursue a more sustainable future' (European Commission, 1995b).

There is currently a credibility deficit at the heart of the European Union's climate change policy which derives from the gap between the rhetoric of leadership and the reality of implementation failure, missed targets and lack of new initiatives. The limited institutional competence of the EU may mean that it will only ever play a supplementary role in the climate policy debate, while more substantive action is expected at the national level. As a collectivity, however, the EU represents a significant voice in global affairs and, perhaps more significantly, is a large contributor to a global problem. It also boasts formidable financial and technological resources and hosts many energy-related industries of global importance such as oil, gas and electricity (Wagner, 1997). And yet the success of future initiatives on climate change is likely to hinge crucially on the extent to which they can be reconciled with wider objectives of the EU in relation to energy security, industrial competitiveness and employment.

6
Water Policy

Introduction

Water policy is one of the oldest and most heavily regulated issue areas in EU (European Union) environmental policy, with a history of legislative activity dating back to the early 1970s. Yet, as this chapter will seek to demonstrate, despite over 25 years of environmental policy activity, the effectiveness of existing measures to control and safeguard aquatic resources in the EU remains open to question. In part, this is because much of the early legislation relating to water policy is out of date and fails to take into account the huge improvements that have been made over the past decades in terms of techniques to control pollution and manage water as a sustainable resource. Doubts have also been raised about the effectiveness of water policy measures in the face of evidence from the European Environment Agency that the quality of water in a large proportion of the EU's rivers and lakes remains below the minimum standards set out in legislation. In the face of growing evidence of regulatory failure and lack of policy effectiveness, in the late 1990s water policy has undergone a process of revision and re-regulation in the EU. This chapter reviews the key developments in the history of water policy in the EU, seeks out evidence of effectiveness (or lack of effectiveness) in environmental policy terms, and assesses the likely impact of recent attempts to re-regulate this area of EU environmental policy.

Public awareness about the environmental consequences of water pollution first came to prominence in the early 1960s. Scientific evidence came to light which indicated that the pesticide DDT (dichloro-diphenyl-trichloro-ethane) could have serious carcinogenic properties when residues were found in food and water after crop spraying. DDT was subsequently banned from use. Public concern about DDT and other sources of water pollution then came to be articulated by Rachel Carson's groundbreaking book, *Silent Spring* (1962), which raised the sceptre of 'rivers of death' resulting from chemical pollution that had 'the power to make our streams fishless and our gardens and woodlands silent and birdless' (Carson, 1962, p. 168). By the end of the 1960s, the public perception that water pollution incidents had become much more common meant that the issue became politically emotive throughout Europe. A clear consensus began to emerge that co-ordinated environmental policies should be introduced to prevent further aquatic pollution. Stimulated by public demand for sufficient quantities of clean and safe water, the European Commission began to formulate environmental policies relating to water in the early 1970s.

Environmental challenges for EU water policy

In EU policy terms, water is divided into various categories. These include fresh water, marine water, groundwater and surface water. Different policies are adopted towards rivers, lakes, estuaries, coastal waters, open sea and underground aquifers. Water is also distinguished by its socioeconomic uses, such as drinking water supplies, water used by agricultural and industry, water used for leisure and tourism and water requiring a particularly high level of conservation.

Yet, although the EU has found it useful to divide water into different categories for administrative purposes, it is important to remember that water itself does not recognise these distinctions. The European Commission has recently acknowledged (European Commission, 1996b, p. 1c) that, in reality, water flows freely between the various categories and often performs a number of functions simultaneously. Due to the natural characteristics of water, it cannot easily be compartmentalised into administrative, policy-motivated, categories.

In EU environmental policy, pollution is defined as

> the direct or indirect introduction as a result of human activity, of substances, vibrations, heat or noise into the air, water or land which may be harmful to human health or the quality of the environment, result in damage to material property, or impair or interfere with amenities and other legitimate uses of the environment. Integrated Pollution Prevention and Control Directive 96/61/EEC)

The EU also sub-divides pollution as being either 'point source' or 'diffuse source' (European Commission, 1996b, p. 3).

In terms of water policy, point source pollution refers to pollutants originating from individual, usually identifiable, discharge points. Point source pollution can include discharges of industrial, domestic or municipal waste water, urban run-off, leakage from storage tanks, industrial installations, farmyards and landfill sites. The environmental damage caused will depend on the nature of the pollution, but may involve potential hazards to human health, detrimental impacts on the ecosystem or disrupting the environmental balance. This pollution may build up in the water over several years, or may occur quickly as the result of an accident that leads to the release of pollutants from a point source into the aquatic environment – for example, from a chemical manufacturing site located near to a water source.

Diffuse source pollution, on the other hand, describes pollution that arrives in water from a number of widely scattered sources that are difficult to identify and control. A typical example of diffuse source pollution is that which results from agricultural practices, including the use of pesticides and nitrates to increase crop yield. Excess amounts of these pesticides and nitrates are then washed from the soil by rainfall and pollute rivers and groundwater intended as drinking water supplies. While the results of diffuse source pollution may well be the same, they normally require different environmental management techniques to reduce or eradicate their effects. Diffuse source pollution may also result from acidification, namely the emission of air-borne pollutants, particularly sulphur dioxide, nitrogen oxide and ammonia from large combustion plant such as coal-fired power generating stations or industrial manufacturing sites. These airborne pollutants can be carried for thousands of miles before being deposited in rivers and lakes through rainfall, resulting in a significantly reduced pH. Although water policy cannot directly address the causes of acidification, EU environmental policy has sought to tackle the acid rain problem at source through the 1988 Large Combustion Plant Directive (88/609/EEC) which sets predetermined targets for the reduction of sulphur dioxide and nitrogen oxide emissions by the member states.

Water may also be polluted as the result of eutrophication (namely, high levels of nutrients in water that lead to an excessive growth of algae at the expense of the natural plant and animal community because the oxygen demand of the algae disrupts the natural balance of the ecosystem), as a result of point source or diffuse pollution, from urban waste water (that is, sewerage) or farming (through agricultural waste and fertilisers).

In addition to environmental problems associated with the quality of water, the availability of water in sufficient quantity is a separate, but nonetheless serious, problem, particularly in southern Europe, where problems of water shortages are more common. Increased demands for water may result in higher levels of abstraction for drinking water, tourism, agriculture or industrial manufacturing. Yet while the quality of water has been the focus of intense EU policy activity, the dangers of over-abstraction have tended to be overlooked. More recently, however, the Commission has been keen to acknowledge that, particularly for groundwater, over-use cannot only lower the water table and damage the aquifer but also lead to the encroachment of salt water into coastal aquifers, causing their loss as a source of drinking or irrigation water (European Commission, 1996b, p. 4). Similarly, over-abstraction from rivers can reduce flow rates that have adverse effects on ecosystems and habitats for plants and animals. Over-abstraction from groundwater or surface waters can also have a range of secondary effects which are nonetheless damaging to the environment, such as the drying out of wetlands or soil erosion.

The physical characteristics of rivers, lakes and coastal areas are often altered by human interference for a range of reasons (see, for instance, European Commission, 1996b, p. 5) including flood protection, canals and waterways, docks and land reclamation. Other human activities – such as leisure and tourism, fisheries and shipping – do not directly seek to alter the aquatic environment, but nonetheless have an impact on it.

Principles of EU water policy

EU environmental policy is underpinned by the principles set out in Article 174 (formerly Article 130r) of the Treaty. These principles have been operationalised by the Commission into a set of interrelated objectives for EU water policy (European Commission, 1996b, p. 5). Accordingly, the Treaty requires that a high level of protection is given to human health, that the 'precautionary principle' be applied. Tim O'Riordan (1992, p. 2) sees the precautionary principle resting on four assumptions: prudent action in advance of scientific certainty; shifting the burden of proof onto the would-be developer to show no unreasonable harm; ensuring that environmental well-being is given legitimate status; and developing best practice techniques in the pursuit of management excellence. In the context of water policy this means that standards are based on recognised scientific knowledge and that a cautious approach is adopted, maintaining higher standards and using the best available techniques wherever there remains scientific uncertainty about the effects on the aquatic environment.

Preventive action which stops environmental damage from occurring is preferred to action which remedies problems once they have occurred. Certainly in the case of water conservation, once a sensitive ecosystem has been destroyed it may be impossible to repair. Preventing pollution at source is also preferable to end-of-pipe solutions so, for example, action which ensures that natural sources of water used for drinking are not contaminated is preferred to expensive treatment to make supplies suitable for human consumption. Following on from preventive action is the principle that environmental damage should be rectified once it has been identified and that the polluter should pay for the cost of measures to repair the damage and discontinue the activity that has caused it. Finally, EU water policy should take account of the principle of sustainable development – namely that environmental concerns should be balanced against socioeconomic factors and the requirement for increased amounts of fresh water to meet demand (European Commission, 1996b, p. 8).

The Commission also recognises that water policy requires coherent integration, both into other EU policy areas and by way of effective implementation of policy at the national and local level (European Commission, 1996b, p. 6). This particularly pertains to the relationship between water policy and agricultural policy, since much of the aquatic pollution that water standards are designed to deal with originates from intensive farming production methods. In practice, the relationship between EU environmental and agricultural policy has often been characterised by competing interests and this is discussed in more detail later in the chapter.

Article 5 (formerly Article 3b) of the Treaty requires that EU action should be taken in accordance with the principle of subsidiarity. This means that water policy measures that can be undertaken most effectively at member state level should not be undertaken at EU level. Even when EU action is taken, subsidiarity also requires that the detailed implementation of water policy should be left to the member states where this is more appropriate. Action at national or local rather than EU level may be considered appropriate because environmental conditions in the EU are likely to vary widely between member states. Water policy that is appropriate in one member state (for example in the UK, where water is relatively fast flowing and contaminants in water are dispersed relatively quickly) may be entirely inappropriate in another (for example, in Spain, where water shortages have been a frequent problem). The Commission therefore applies the principle of flexibility to ensure that the most appropriate policy is implemented in a particular region (European Commission, 1996b, p. 7). However, it is also the case that water pollution does not observe national boundaries. It may well have impacts across a number of member states. Where there is potential for transfrontier pollution, there is often sufficient justification for the EU to act (European Commission, 1996b, p. 8).

This chapter will now outline the development of EU water policy in order to put more recent initiatives into historical perspective. EU water policy has developed in three distinct phases, each of which are described below.

Origins of EU water policy

From a public policy perspective, one of the key reasons why water quality emerged at the forefront of the EU's environmental policy agenda is undoubtedly the fact that water supply and water treatment was, and remains, a largely publicly owned sector throughout Europe. Reaching consensus on environmental policy objectives and on the precise form of policy instruments is undoubtedly more straightforward when the private interests of corporate actors do not have to be taken into account. The water supply and treatment industries were, and largely remain, much weaker in EU lobbying terms than other industrial sectors that represent predominantly private sector companies.

The relatively weak EU lobbying position of water supply and treatment companies is only partly due to structural problems in the industry which have made it difficult to achieve consensus through their European federation, Eureau, on a strategy of opposing higher environmental standards. To a large extent, water companies have avoided outright opposition to improved water quality standards because they too recognise the need to improve the quality of the services they provide. It would, in any event, be unpalatable in public relations terms to oppose improvements and, in any case, the additional costs of complying with higher environmental standards can largely be passed onto consumers in their water bills rather than being a cost to be internalised by the companies themselves. Unlike other industry sectors, water consumers are generally unable to seek out another supplier (that is, a non-EU supplier) if prices rise steeply due to environmental compliance costs. The absence of a price-sensitive market for water supply and treatment removes much of the stimulus that might otherwise have existed for the industry to lobby against higher water quality standards in EU environmental legislation, or at least to question the validity of the scientific evidence on which EU environmental standards have been based.

In contrast to the relatively low profile adopted by the water industry in lobbying terms, environmental non-governmental organisations (NGOs) have become increasingly skilled in presenting scientific evidence to justify the need for higher water quality standards in EU legislation. Friends of the Earth, in particular, has successfully used scientific evidence of aquatic pollution to paint a picture of the 'worst-case' scenario that would result if EU environmental legislation were not put in place. Friends of the Earth has also has been active in bringing cases before national courts in the UK, claiming the government has failed in its duty to implement EU water Directives (see, for example, *R.* v. *Secretary of State for the Environment ex parte Friends of the Earth and another*, Court of Appeal, Civil Division, 25 May 1995).

For Richardson (1994), the result of uncertain agendas, shifting networks and complex coalitions in EU water policy is that not only is EU legislation very extensive in scope, but it also has major cost implications for the member states. Jordan (1999b), furthermore, has argued that once water policy standards are embedded in national political systems via EU Directives and become extremely difficult to reform, even when they create enormous political problems for member states and are outdated scientifically. From a public policy perspective, therefore, member states appear locked into a sub-optimal trajectory (Jordan, 1999b, p. 13).

But aside from purely public policy explanations, water also came to the fore of EU legislative activity for practical reasons: aquatic pollution was, at least initially, a more tangible form of degradation than other environmental incidents such as those affecting air or soil quality. Coastal and river pollution could not be ignored. Its effects were highly visible, as marine life suffered from the intrusion of human activities, and it held no respect for national boundaries, with pollution incidents in major rivers such as the Rhine leading to environmental damage in a number of member states. From the 1960s onwards, the case for a coordinated environmental policy on water grew ever stronger.

Water policy after the First Action Programme on the Environment, 1973

In 1973, the First Action Programme on the Environment (European Community, 1973) identified water pollution as an issue where priority action was required (see Bell, 1997, p. 439) and the earliest water quality legislation was adopted by the Council in 1975. Due to the public perception that ever higher water quality standards are required to ensure public health and prevent further environmental degradation, and coupled with the absence of private corporate interests to present a contrary argument based on environmental and cost efficiency, legislative activity was rapid. There are now over 20 Directives or Decisions that deal directly with water policy or are closely related to it (Haigh, 1995, 4.2–1). The most important of these are described below, in chronological order:

The Surface Water Directive (75/440/EEC) had the objective of ensuring clean drinking water sources by requiring member states to identify, classify and set up action plans to ensure that quality standards were achieved by rivers, lakes and reservoirs used as drinking water sources. The standards in the Directive are now out of date and, in any case, have now been superseded by the Drinking Water Directive. The Surface Water Directive has also been criticised (see, for example European Commission, 1996b, p. 20) because its value in protecting sources of drinking water remains unproved.

The Bathing Water Directive (76/160/EEC) seeks to safeguard the health of bathers and maintain the quality of bathing waters by requiring member states to identify marine and fresh water bathing waters, monitor them and take 'all appropriate measures' to ensure compliance with quality standards. The Commission considers the Bathing Water Directive to be very popular with EU citizens (European Commission, 1996b, p. 21). However, there is a popular misconception that it is the Directive that results in the award of 'Blue Flags' for clean beaches. These are actually awarded by the completely unrelated Foundation for Environmental Education in Europe. Nevertheless, at the time the Directive was adopted, there was little other EU legislation on water pollution. The Bathing Water Directive was the first attempt to deal with pollution problems caused by the disposal of sewerage and waste water although, more recently, the Urban Waste Water Directive has dealt directly with this issue. In 1994 the Commission published a new proposal to update the Bathing Water Directive.

EU water legislation has tended to focus on point source

pollution control. This is simply because it is one of the easiest problems to recognise and take action against (European Commission, 1996b, p. 9). Although member states have retained discretion over how to implement the detail of EU water legislation, the basic principle has been to require economic activities (such as industry) to be licensed by the competent national authorities and to make the granting of that licence dependent on pollution control measures being put in place. These pollution control measures are normally in terms of emission controls on the amounts of pollution that may be discharged into water and quality standards that rely on assessments of the overall quality of the aquatic environment. The resulting tension between the emission controls and quality standards approaches is exemplified by the history of the Dangerous Substances Directive.

In an attempt to control surface water pollution, the Dangerous Substances Directive (76/464/EEC) required member states to control emissions of all dangerous substances listed in the annex of the Directive. The main control mechanisms are permits, issued to industrial installations and by improved treatment of urban waste water. The conditions under which permits may be issued for the more dangerous substances (set out in List I of the annex) were to be laid down in 'daughter Directives'. Member states then had a choice between two methods for setting these conditions, either an emission 'limit values' approach or on limits required to meet specified 'quality objectives' in the water receiving pollution. Less dangerous (List II) substances are subject to pollution reduction programmes in each member state. The emission limit approach is based on estimates of the maximum level of reductions in pollution that could reasonably be expected given the best available techniques not involving excessive costs (BATNEEC). The meanings of 'best available techniques' and 'excessive costs' have often proved difficult to ascertain. On the basis of these estimates, emissions are allowed under the terms of each licence issued. This approach is widely used in the United Kingdom, where the Environment Agency operates a discharge consent system (Bell, 1997, p. 440). In other member states, the quality standards approach tends to be used as the basis for issuing pollution control licences. The quality standards approach seeks to establish the quality objectives that are to be achieved, then estimate the amount of pollution that water is likely to tolerate without harming the aquatic environment. Permits are then issued to potential polluters, based on the geographical areas where the quality standards would apply.

The fact that water Directives have allowed member states to choose between their preferred approach of pollution control can be attributed to a political compromise that resulted in both the quality standards and the emission limits approaches being retained (Somsen, 1990, p. 93). This compromise is not considered ideal (Bell, 1997, p. 440; Haigh, 1995, 4.2–1). In practice, the Commission recognises that neither of the approaches offers an ideal solution because, while emission limits can lead to unnecessary investment without significant benefit to the environment (Matthews and Pickering, 1997, p. 265), the quality standards approach can be abused as a 'licence to pollute' up to a defined level (European Commission, 1996b, p. 10). Most water Directives nevertheless continue to persist with elements of both the emission limits and the quality standards approaches.

However, although the Dangerous Substances Directive has played an important part in improving surface water quality, the Commission has recognised that the procedure for producing 'daughter' Directives for List I substances has proved burdensome and slow, while few member states have taken action to reduce pollution from List II substances at all (European Commission, 1996b, p. 22). The problem that member states have had in implementing the Directive is that the list of potentially dangerous substances continues to grow. Also, because the daughter directives deal with each substance individually, they do not consider the potential toxicological effect of a 'cocktail' of these substances mixed together in water. An attempt to answer these criticisms has been made in the recent Integrated Pollution Prevention and Control (IPPC) Directive.

After the 1976 Dangerous Substances Directive, the next significant policy measure proved less controversial. This was the Information Exchange Decision (77/795/EEC), which set up a network of 124 monitoring points (in 12 member states) to measure the quality and quantity of water according to 19 criteria. This information is then exchanged between the member states and published by the Commission. The Decision was subsequently superseded by the activities of the European Environment Agency (EEA) and by the monitoring and reporting requirements of later Directives, but the Commission's view is that the Decision has proved useful in providing a long time series of data on the quality and quantity of water in the EU (European Commission, 1996b, p. 22).

The Fish Water (78/659/EEC) and Shellfish Water (79/923/EEC) Directives have the objective of protecting fresh water capable of supporting fish life, and coastal and brackish waters that support shellfish, from pollution. The Directives require member states to designate fish and shellfish waters, establish quality standards, monitor these waters and reduce pollution levels. Since member states have complete discretion over which waters are to be covered by the Directives, they have been implemented differently and have been criticised for having a patchy impact across the EU.

The Groundwater Directive (80/68/EEC), which started life as a companion to the Dangerous Substances Directive, seeks to prevent dangerous substances from polluting groundwater. However, the Commission acknowledges that, since pollution frequently comes from diffuse, not point, pollution sources (such as the use of agricultural pesticides), and because there is a separate problem of over-abstraction, the Directive does not adequately address the environmental problems for groundwater (European Commission, 1996b, p. 24).

In order to ensure that water intended for human consumption is safe, the Drinking Water Directive (80/778/EEC) sets quality standards for more than 60 parameters. Compliance with these parameters is monitored by the member states, who report to the Commission. The Directive has led to investment in water filtration plant by water supply companies in the member states. It has improved the quality of drinking water that consumers can expect to receive from their taps, but often with a high cost reflected by increased charges in water bills (see Matthews and Pickering, 1997, p. 265). In response to these criticisms, in 1995 the Commission published a proposal to revise and update the standards laid down in the Drinking Water Directive which have now been adopted.

Revision of water policy after the Frankfurt Ministerial Seminar, 1988

Despite the legislative activity that took place after 1973, the Commission was adopting a rather piecemeal approach to water policy. Not all water pollutants were even covered by the legislation (Bell, 1997, p. 439). In 1988, when a ministerial seminar was held in Frankfurt to review progress made with regard to water policy, the meeting recognised that significant improvements needed to be made to the existing body of water legislation. Subsequently, additional legislation was adopted on urban waste water treatment and nitrates.

The Urban Waste Water Treatment Directive (91/271/EEC) seeks to reduce pollution by setting conditions for the treatment and discharge of urban waste water (sewerage) and from waste water from industrial sectors. The Urban Waste Water Treatment Directive combines the quality objectives and the emission limit approaches to pollution control. It is in the process of being implemented by the member states and, as with the Drinking Water Directive, is expected to result in large increases in water prices for consumers.

The Nitrates Directive (91/676/EEC) seeks to complement the Urban Waste Water Directive, by controlling nitrate pollution from agricultural sources. It requires member states to produce and promote Codes of Good Agricultural Practice to reduce the level of nitrate loss to surface water and groundwater from agriculture, and to monitor areas identified as being vulnerable to nitrate pollution. As with the Urban Waste Water Directive, the Nitrates Directive combines the quality objectives and emission limit approaches to pollution control.

A further ministerial meeting on water policy was held in The Hague in November 1991, resulting in a Council Resolution on the future of groundwater policy (European Communities, 1992a, p. 2). The Resolution called on the Commission to draw up a detailed programme for the protection and management of groundwater. A draft Groundwater Action Programme was subsequently published in 1996 (European Communities, 1996).

To complement the groundwater proposals, in 1993 the Commission proposed a further directive on the Ecological Quality of Water Directive (COM (93) 68) which seeks to maintain and improve the quality of surface waters by requiring member states to monitor the ecological status, identifying potential sources of pollution, and establish targets and programmes for achieving water of good ecological quality.

Also in 1993, a detailed statement on the status of EU water policy was made in the Fifth Action Programme on the Environment. It reviewed progress made by water policy in the light of the fact that water as one of the elementary sources of life, an indicator of the general quality of the natural environment, and a prerequisite for a harmonious and sustainable development of socioeconomic activities (European Communities, 1993, p. 50). The Fifth Action Programme also acknowledged a principle not apparent in all earlier water policy decisions, namely that EU policy should take into account not only the quality of water available, but also ensure that it is available in sufficient quantities to achieve sustainable development without upsetting the natural equilibrium of the environment.

Accordingly, the Fifth Action Programme on the Environment (European Communities, 1993, p. 51) set out three policy aims in relation to the management of water resources:

- prevention of pollution of fresh and marine surface waters and groundwater, with particular emphasis on prevention at source;
- restoration of natural ground and surface waters to an ecologically sound condition, thus ensuring (inter alia) a suitable source for extraction of drinking waters;
- ensuring that water demand and water supply are brought *into equilibrium* on the basis of more rational use and management of water resources.

Moreover, the Fifth Action Programme recognised the importance of water as a sustainable resource for a number of economic sectors, including industry (which uses water in manufacturing processes), the energy sector (which uses water as a coolant in generating processes), the agricultural sector (which particularly uses water for crop irrigation) and tourism (which relies on the provision of clean and safe drinking and bathing waters). However, the Fifth Action Programme also acknowledged that these sectors are the main contributors to water pollution and the main cause of over-abstraction.

In line with The Hague Declaration of 1991, the Fifth Action Programme also set out the objectives for water quantity and quality until the year 2000 (European Communities, 1993, pp. 53–4). For groundwater and surface water, the quantity objectives were for the sustainable use of fresh water resources which balances demand with availability. The objectives for the quality of groundwater were to maintain uncontaminated groundwater, to prevent further contamination of polluted groundwater, and the restoration of contaminated groundwater to drinking water quality. The objective for the quality of fresh surface water was to maintain a high standard of ecological quality with a high level of biodiversity. Finally, the objective for marine water quality was the reduction of discharges of toxic substances which, due to their persistence or accumulating impact, could negatively affect the environment.

At the Environment Council meeting on 20–23 June 1995, member states called on the Commission to undertake a much more thorough review of water policy in the light of the objectives set out in the Fifth Action Programme. The Commission published its response on 21 February 1996 in a Communication to the Council and the European Parliament. In its Communication (European Commission, 1996b, p. 2), the Commission elaborated on the objectives of sustainable water policy as being: to provide a secure supply of drinking water that is safe and available in sufficient quantity and with sufficient reliability; to provide water of sufficient quality and quantity to meet drinking and other economic requirements (that is, for industry and agriculture and to sustain fisheries, transport and power generation activities as well as meeting recreational needs); to ensure that the quality and quantity of water is sufficient to protect and sustain the good ecological state and functioning of the aquatic environment; and to ensure that water is managed so as to prevent or reduce the adverse impact of floods and minimise the impact of droughts.

The Commission acknowledged that the four objectives of EU water policy would not always be mutually compatible. A sustainable water policy was seen as being one that achieves a balance between objectives but, overall, the protection of quality and quantity of water resources was considered the priority for EU policy. To achieve the objectives of a sustainable water policy, the Communication recommended the adoption of a framework Directive. This marked the beginning of a process of consultations between the Commission and the Council, European Parliament, Economic and Social Committee and the Committee of the Regions. The Commission also received written submissions from environmental and consumer NGOs, water supply companies and national environment agencies and, on 28–29 May 1996, held a conference on the future of EU water policy at which interested parties were invited to participate.

A new approach to water policy after 1997

The outcome of the Commission's consultations was the publication, on 26 February 1997, of a proposal for a Directive establishing a framework for action in the field of water policy (European Commission, 1997a). The Commission later made some technical amendments to the proposal in November 1997 (European Commission, 1997d) and February 1998 (European Commission, 1998a). The proposed legal base of the framework Directive is Article 175(1) (formerly Article 130s(1)) of the Treaty and it seems likely to be adopted as legislation under the cooperation procedure.

The justification for a new approach to EU water policy is that much of the existing body of legislation, dating back to 1975, is now out of date (see also Matthews, 1998). The existing body of legislation has been criticised by the European Parliament (1996, p 52) as being, in many respects, contradictory since earlier legislation has since been superseded by subsequent measures without actually being repealed. It is also needs to take into account the objectives of a sustainable water policy set out by the Commission in 1996 (European Commission, 1996b). Certainly, water policy over the past 25 years cannot be judged to have been a total success. Despite a quarter of a century of EU legislative activity designed to ensure good quality water, in 1994 the European Environment Agency reported that only 10 per cent of water in the EU's rivers and lakes met the EEA's criteria for good quality (European Environment Agency 1994, p. 48). This reconfirmed the findings of the European Environment Agency's Dobris Report (European Environment Agency, 1992) which had found that there was still much to be achieved to protect the aquatic environment of Europe.

The framework Directive on water policy rationalises much of the earlier water legislation and responds to a number of issues not previously dealt with by EU legislation, particularly over-abstraction and water shortages. Once it has come fully into force, much of the existing water legislation will be repealed and set out an overall framework into which the remaining directives – including the Bathing Water, Drinking Water, Urban Waste Water Treatment and Nitrates Directives – would fit. Once the framework Directive is fully operational, the intention is that the EU will repeal the Surface Water Directive, the Information Exchange Decision, the Fish Water Directive, the Shellfish Water Directive and the Groundwater Directive, while the Commission's proposal for a Directive on the Ecological Quality of Water would be withdrawn and, in effect, extended to a broader concept of integrated water policy (European Commission, 1996b, p. 14).

The framework Directive also constitutes a new approach for EU water policy because, in contrast to earlier water directives, it provides for an integrated framework to protect surface water, groundwater, estuaries and coastal waters together. It is based on the principle that EU policy should recognise the fact that water is not static. It flows from rivers into groundwater, lakes and the sea. In this respect, the framework Directive follows an approach to environmental management developed in the United States. There have been some doubts that it can provide an appropriate model for the very different environmental conditions experienced in Europe (European Parliament, 1996, p. 14) since diffusion of pollution might be considered easier to achieve in the US than in the geographical and environmental terrain of the European Union.

The framework Directive creates a structure within which the EU, national, regional and local authorities can develop an integrated approach towards water policy. It seeks to encourage cooperation between different member states in order to identify policy areas where further action is required and to take more effective action on water policy, particularly where river basins cross national frontiers, such as the Rhine and the Danube (European Commission, 1996b, p. 16).

The framework Directive has four objectives: the provision of drinking water; the provision of water for other economic activities; the protection of the environment; and the alleviation of the impact of floods and droughts. The Commission has stated that protection of the environment is the Directive's main objective, while the prevention and alleviation of floods and droughts will not be achieved by the framework Directive alone, but will be a general objective of EU water policy (European Commission, 1997a, p. 5).

The main tool that the framework Directive uses to coordinate implementation of the Directive is the river basin management plan. A system of river basin management districts will be established in each member state, with each district including more than one river basin and providing the primary administrative unit for coordinating implementation of the Directive. Groundwaters and coastal waters are to be assigned to the 'nearest or most appropriate' river basin. Designated competent national authorities will prepare a review of each river basin, including an analysis of the characteristics of river basins, economic analysis of water use, environmental assessments and the impacts of human activities on water resources.

The competent national authorities will then be responsible for consulting with interested parties and the public before implementing the river basin management plans. The management plans are required to specify what programmes will be implemented to achieve the environmental objectives for each basin, including a review of the analysis undertaken, monitoring data and an indication of the measures that will be taken to meet environmental objectives within a specified timetable, with the intention that all waters in a river basin achieve 'good' status.

In terms of improving public access to environmental information, the framework Directive is a considerable improvement on earlier policy. River basin management plans will be subject to public consultation when they are in draft form, followed by the publication of each plan one year before it comes into operation, with at least six months allowed for the public to make comments. River basin management plans are due to be put in place ten years after the Directive comes into force and to be fully operational six years later. Each management plan must then be updated every six years after that. This timetable is much more generous than the one originally proposed by the Commission when the Directive was conceived. This reflects the desire of the Council to leave longer transitional periods during which national procedures and administrative practices can be adapted. Once the management plans are operational, each member state will send copies to the Commission and the European Environment Agency. This will mean that, for the first time, the EU will have information on Europe's aquatic environment in a standardised format which can be easily analysed and compared.

River basin management plans will include programmes of action defined as either 'basic measures' or 'supplementary measures'. Basic measures include instruments to achieve environmental quality standards, to implement other relevant legislation and to introduce water charges. Supplementary measures include instruments necessary to meet other objectives of the framework Directive, particularly in terms of sustainable quantities of water consumption. When water fails to meet 'good' status, member states will be required to take other interim measures, including additional monitoring, investigation of likely sources of pollution and a review of pollution and discharge permits.

The idea of river basin management plans is an innovative approach for water policy that builds upon and replaces the Commission's earlier proposals for a Groundwater Action Programme (COM(93) 680 final). As such, the framework Directive marks a new integrated approach to water pollution control and demand management that is in marked contrast to earlier legislation which treated different aspects of water policy separately.

The main environmental objective of the framework Directive is that, by the end of the transitional period, member states will ensure that surface water and groundwaters in each river basin district are of 'good' status. The inclusion of a deadline by which member states must meet this objective overcomes many of the criticisms about the earlier Commission proposal for a Directive on the Ecological Quality of Water, which would have left member states free to determine for themselves the surface waters to be improved. Member states will, however, be allowed additional time if natural conditions do not allow for rapid improvements in water quality. It is possible that, in some member states, the

Directive will not be fully operational until 34 years after the legislation was adopted (ENDS, 1998b, no. 281, p. 47). Furthermore, where river basins have been severely impacted by earlier human activities and improvements in status prove to be impossible or prohibitively expensive to achieve, the Directive allows for longer time scales still. One major concern is that, since the Directive does not define what is meant by 'prohibitively expensive', it will be left to the discretion of member states to decide with improvements to water quality are too costly to achieve. There are similar concerns that the Directive does not define what is meant by 'good' surface waters or groundwaters. This lack of clear definitions is partly deliberate since the framework Directive does not reconcile the emission limit and quality standards approach found in earlier water Directives. Standards set by the daughter Directives to the Dangerous Substances Directive, for instance, would be incorporated into the framework Directive on water quality as part of the criteria used to describe water of 'good' chemical status. However, because relatively few 'daughter' Directives have been put in place for individual chemical substances, many dangerous substances will not be taken into account at all.

The definition of 'good' surface waters also lacks elaboration in the framework Directive. It awaits future discussion by a scientific management committee, outlined below, which will consider the chemical and ecological status of surface waters. Groundwaters, on the other hand, will be assessed in relation to their chemical and ecological status. The overall status of surface waters and groundwaters will also be taken into account and this, in turn, will be based on the least well achieved of the chemical and ecological criteria. Defining 'good' ecological status is particularly difficult. Despite the Commission's proposal that 'good ecological status shall entail the achievement of any physico-chemical, physical and biological standards' (European Commission, 1997d, p. 15), there is a distinct lack of reliable technical information that will hamper the task of elaborating what good ecological status actually means.

The key to the Commission's new approach to water policy, and the means by which water of 'good' status will be defined, is through a management committee (European Commission,

1997a, p. 45). This will be made up of scientific experts from each member state and chaired by a Commission official. The reason why so much of the detail of water policy will be left to a non-elected committee of technical experts is that state-of-the-art scientific standards need to be ensured. In the past, water policy has been criticised for being out of date in the light of newer toxicological data (see, for instance, Matthews and Pickering, 1997, p. 265). By introducing a flexible means by which EU environmental standards for water can reflect advancements in scientific knowledge, the framework Directive answers many of these earlier criticisms.

However, by leaving so much of the detail of EU water policy to a management committee which meets only after the framework Directive has been agreed by the member states, a great deal of power is devolved to a new committee structure. In practice, the role of a scientific management committee in setting water policy makes good sense but, in political terms, several member states remained concerned that the new structure will lead to more policy being determined at EU level, ultimately resulting in an erosion of national sovereignty. During early negotiations for the framework Directive, the UK was particularly concerned about the adverse impact on national sovereignty, arguing that rather than delegating so much detail to a management committee, definitions of 'good' status water should have been agreed by the member states before the Directive was adopted. In the event, this did not happen since it would have required the Council of Ministers to engage in complex and detailed negotiations on technical standards which, conceivably, could have been out of date soon after the legislation was adopted. The flexible approach that relies on a scientific committee structure subsequently prevailed, but concerns still remain that there will be a lack of public access to its decision-making procedures.

The framework Directive on water policy is also a major development because it allows member states to use economic instruments, often known as 'green taxation', in order to achieve EU water policy objectives (European Commission, 1997a, p. 38). This is only the second time that economic instruments have been used for EU environmental policy, the first being the ill-fated proposals for an EU energy tax (discussed in Chapter 5).

The aim is to prevent the over-abstraction of water by ensuring that charges for water use cover the full costs to the environment. Higher water charges are intended to encourage more efficient use of water and ensure that the environmental costs of water use are borne by those using it. This enshrines elements of the 'polluter pays' principle and the aim of 'sustainable development' that are central to EU environmental policy. It is left to the discretion of each member state to decide what charges for water use are appropriate, but such charges are likely to be particularly costly for agricultural and industrial users, although domestic users may also be asked to pay more for the water they use.

This measure is potentially the most expensive aspect of EU water policy ever conceived, with the resulting increases in water bills even larger than those resulting from the Drinking Water Directive and the Urban Waste Water Directive. In practice, the actual impact on water prices will depend on how individual member states choose to balance the need to ensure that water remains a sustainable resource with a desire to keep water prices at an affordable level. This will require delicate political and economic decisions on the part of national governments. In the UK, where the water industry is privatised, the Commission has indicated that the principle of full cost recovery for the use of water is already applied in practice but in other member states, where water prices remain artificially low, the implications of higher prices may be more far-reaching. Although the idea of using economic instruments to achieve environmental objectives is widely acknowledged to be sound, and is certainly in line with the 'polluter pays' principle outlined in Article 174 (formerly Article 130r) of the Treaty, hard political decisions still need to be taken at the national level about exactly who should pay the environmental costs of water use. There have also been some criticisms that the framework Directive relies too much on end-of-pipe solutions and that plans for economic instruments do not take sufficient account of the 'precautionary principle' by preventing pollution from happening in the first place. The costs of cleaning up diffuse source agricultural pollution from pesticide use, for example, are largely paid for by domestic water users and not by the farmers who are responsible for using increasing amounts of organochlorines to obtain higher crop yields in the first place.

The framework Directive on water policy is an important measure for updating and rationalising earlier EU environmental legislation. Nevertheless, it retains the distinction between emission limit and the quality standards approach that was first adopted as a compromise in the Dangerous Substances Directive. By retaining the quality standards approach favoured by the UK alongside the emission limit approach used elsewhere in the EU, the framework Directive falls short of an entirely new approach for water policy. The European Parliament has been particularly critical of the failure to resolve the compromise of a dual approach to water quality standards, or even to provide a clear indication of how the correct balance between the two systems should be made (European Parliament, 1996, p. 52). The framework Directive thus falls short of an entirely new and or fully integrated water policy.

Instead, the framework Directive replaces some earlier measures but retains others. The Drinking Water, Bathing Water and Urban Waste Water Directives, for instance, are retained and there are some concerns that this new framework for EU water policy will not full integrate all measures within a common structure, despite this being one of the original objectives of a new approach. There also remains a lack of integration between water policy and other EU policy areas, particularly agriculture. The tension between water and agricultural policy is not new (Matthews and Pickering, 1997, p. 270) and improved agricultural practices to tackle diffuse pollution are still needed in order to take account of the 'polluter pays' principle. For instance, the tortuous progress of the draft Groundwater Action Programme, a Commission initiative to prevent pollution (particularly by agricultural pollutants) at source, can be attributed to the fact that DG VI (the Directorate-General for Agriculture) viewed this initiative as an attempt by DG XI (Environment) to introduce measures that would impact on a key area of its own policy competence – namely the operation of the Common Agricultural Policy (CAP). The draft Groundwater Action Programme has now been incorporated into the framework Directive on water policy, although it remains to be seen whether the division of policy competence between the Agriculture and Environment Directorates-General of the Commission will continue to be the source of a continued policy impasse for initiatives to curb agricultural pollution of the EU's aquatic environment in the future.

One important aspect of the framework Directive is the integration of water quantity and water quality aspects of EU legislation. Earlier Directives focused almost entirely on water quality and did not take sufficient account of the fact that water is a finite resource that must be replaced by new stocks if abstraction and usage is to be in line with the principle of sustainable development. One criticism of the sustainable approach to water policy taken in the framework Directive, however, is that it does not take account of climate change and the possible impact on the quality and quantity of water available through drought or flooding.

Conclusion

The new framework approach to water policy carries with it the recognition that much of the earlier body of EU legislation failed to achieve its environmental objectives, was burdensome and difficult to apply in practice. The framework approach simplifies existing water Directives, replacing overly prescriptive standards with a single set of broad environmental quality and quantity objectives. However, where the framework Directive is lacking is outside the sphere of environmental policy per se. Its failure to integrate other policy areas, particularly agricultural policy, into its framework limits the potential for a coherent strategy of sustainable development in the European Union. The decision to retain some earlier water legislation, particularly the Bathing Water, Urban Waste Water and Drinking Water Directives, also carries with it the potential for policy conflict. The Drinking Water Directive, for instance, still retains fixed maximum admissible concentrations (MACs) for substances that are permitted in water supplies. These fixed MACs for drinking water seem at odds with the framework approach – with its emphasis on flexibility and state-of-the-art standards for EU water policy. Nevertheless, in that the framework Directive on water policy endeavours to take account of best available techniques and best environmental practices, it is a marked improvement on the inflexible, often out-of-date, environmental standards set in earlier EU legislation. It signals a move away from prescriptive, standard-setting legislation and offers an innovative and responsive approach to water policy for the European Union.

7
Air Pollution

Introduction

The traditional emphasis of air pollution policy was on 'stationary' sources of air pollution: either on large industrial plants which poured out smoke and substances harmful to health, or on the effects of coal fires in domestic households which produced 'smogs' in cities such as London which seriously impaired movement and led to significant numbers of additional deaths through respiratory diseases. The number of industrial plants has declined throughout Europe, and those that remain use more efficient methods for reducing pollution. Domestic legislation created 'smokeless zones' and, in any case, the spread of central heating and other new forms of space heating reduced the domestic use of coal. Although some stationary source problems remain, the focus of the debate about air quality management in recent years has been on pollution from 'mobile sources' – such as cars and trucks – and the consequent hazards to human health. For example, in London traffic contributes 97 per cent of carbon monoxide emissions, 77 per cent of particulate matter and 75 per cent of nitrogen oxides (*Air Quality Management*, March 1998, p. 1). This chapter will focus on mobile source problems which have been the principal policy concern of the EU in the late 1990s.

This is an area which faces particularly difficult problems of policy formulation and implementation. One cannot easily satisfy the fourth criterion of effectiveness, that of changing the behaviour of relevant actors. The pollution is largely caused by a very large number of individual citizens exercising what they would see as their right to drive a motor vehicle where they want to when they want to. In so doing, they are consuming products made by two major industries, the oil industry and the motor vehicle industry. Any drastic measures to cut air pollution from vehicles is therefore likely to face opposition from a coalition of large numbers of citizens and two powerful industries. Hence the attractiveness of technological solutions which allow everyone to maintain their existing patterns of production and consumption.

Air pollution and health

Although air pollution may have effects on visibility, the decay of urban buildings and vegetation, and possibly on weather conditions, the principal area of concern is its effect on human health. The fifth condition of effectiveness noted in the introduction to this book was that policy remedies should enjoy intellectual integrity. There are, however, a number of uncertainties associated with the relationship between air pollution and health to which medical experts are not always able to provide a clear answer. In order to understand policy, it is necessary to have some understanding of the main pollutants and their possible effects.

A fundamental methodological problem is estimating how long victims would have lived if they had not been exposed to pollution. Many such deaths occur among elderly people with respiratory illnesses who have limited life expectancy. However, in Britain an expert panel of the Department of Health, the Committee on the Medical Effects of Air Pollution, has estimated that between 12 000 and 24 000 deaths are brought on each year in the UK by poor air quality.

One particular area of concern has been the extent to which the increased incidence of asthma is related to poorer air quality resulting from increased traffic volumes. Once again, there are methodological difficulties: the growth in the volume of road traffic has coincided with the growth in double glazing, central heating and wall-to-wall carpeting. The consequent warm, humid and controlled environment in turn has favoured the reproduction of the household dust mite, the major allergen associated with the disease in Britain. The importance of indoor conditions is reflected in the fact that asthma prevalence is as high if not higher in the largely traffic-free Scottish islands as in many UK urban areas. Following the publication of the report from the Committee on the Medical Effects of Air Pollution on asthma and outdoor air pollution it was observed that

> Asthma is clearly a disease in which many factors [operate], and a public obsession with one [air pollution], obscures others which individuals may be much more able to influence. Avoiding exposure to tobacco smoke, furry pets or house dust mite during infancy may reduce the chances of developing asthma. (*Air Health Strategy*, January 1996, p. 9)

It is also important to distinguish between air pollution as a possible cause of asthma in genetically susceptible individuals and its more likely role in triggering attacks where it 'may be the most important trigger for attacks in some people' (*Air Health Strategy*, January 1996, p. 9).

In many respects, the jury is still out on the link between pollution and asthma. 'There is little or no association between the regional distribution of asthma and that of air pollution' and studies 'have not found consistent associations between outdoor air pollution and asthma prevalence' (*Air Quality Management*, June 1999, p. 8). The difficulties over asthma reflect more general methodological problems in exploring the links between health and air pollution. Many studies either contradict each other or make claims on the basis of relatively small samples.

In order to understand the health problems arising from air pollution, it is necessary to distinguish between each of the principal air pollutants and identify their particular health effects. Scientific knowledge on these matters is continually advancing, and in some areas is running ahead of the capacity of policy-makers to respond, as will be noted below in relation to the issue of fine particles. The task of policy-makers is not made easier by frequent disagreements among medical experts about the nature, extent and seriousness of the health effects of particular pollutants, often arising from the methodological difficulty of separating out the influence of a range of factors on human health. The options available to policy-makers are further limited by interactive effects among some of the pollutants and the considerable distances over which some pollutants can travel.

Ozone in the lower atmosphere has been regarded as one of the pollutants most harmful to health, although in recent years increasing attention has been focussed on fine particles. Ozone is a secondary pollutant which forms as a result of a photochemical reaction in sunlight between nitrogen oxides and volatile organic compounds. The principal source of nitrogen oxides (NOx) is from road transport, although emissions in Britain peaked in 1989. According to the EEA, around 40 per cent of Europe's population lives in cities with average nitrogen dioxide concentrations above the EU's guide value (*Air Quality Management*, August 1997, p. 7). Volatile organic compounds come from a range of sources, such as paints and solvents. 'Some of the highest concentrations of ozone are found in rural areas where photochemical reactions have had time to occur in polluted air dispersing from urban and industrial centres' (House of Lords, 1996, p. 8).

The lungs are the main target of ozone. For healthy people, it can make breathing more difficult and chronic exposure can damage deep portions of the lung, although levels encountered in the UK are unlikely to have such an acute effect. Recent evidence from the United States suggests that long-term exposure to ozone pollution may permanently damage the lungs (*Air Quality Management*, July 1998, p. 7). Its main threat, however, is to those already suffering from existing respiratory diseases such as emphysema, bronchitis and asthma. For asthma sufferers, 'ozone sensitises people with asthma to common allergens such as pollen by damaging the ability of the cells which line the respiratory tract to sweep out invading pollen and bacteria' (Elsom, 1996, p. 62).

Sulphur dioxide originates principally from power stations and heavy industry and major reductions in emissions have been achieved since 1970. It can constrict the airways and is therefore a particular health hazard for asthmatics. In some parts of the UK, 'levels of sulphur dioxide regularly exceed those at which effects of clinical significance, including tightness of the chest, coughing and wheezing have been demonstrated in [asthmatic] individuals' (Committee on the Medical Effects of Air Pollutants, 1994, p. 16).

Carbon monoxide is the result of incomplete combustion of motor vehicle exhausts. Emissions are particularly high when an engine is started from cold or it is idling. It is therefore a particular problem in areas with congested or high volumes of traffic and during the winter months. Carbon monoxide is readily absorbed into the body from the lungs, leading to reduced levels of oxygen reaching body tissues. Even healthy people exposed to high concentrations may experience headaches, fatigue and slow reflexes. It may be harmful to those suffering from heart disease such as angina. For drivers, keeping windows and air vents closed can significantly reduce expose to carbon monoxide (Clifford, Clarke and Riffar, 1997).

In the last few years, increasing attention has been given to the health effects of particulate matter (PM). This is 'the generic term given to very small particles present in the atmosphere. Recent research indicates that episodes of high atmospheric PM concentrations correlate to increases in asthma attacks and deaths from respiratory illnesses' (House of Lords, 1996, p. 9). Unlike ozone, which is an invisible gas, 'The public can see particle pollution, or rather the black smoke that makes up some of the particle pollution. The fact that they can see it, smell it, understand it and blame someone else (i.e. trucks and buses) for its causation leaves an obvious popular scapegoat' (*Air Quality Management*, June 1999, p. 9).

Unlike a gas of uniform composition like ozone, particles vary in their size and composition. Fine particulates are ten micrometres or less in diameter (PM10) and hence invisible to the naked eye. The bigger a particulate, the potentially less harmful it is. Such larger particles are often windblown dusts and other 'natural' substances. They settle to the ground relatively quickly and if one of them is inhaled, it will tend to collect in the throat or nose and be eliminated from the body easily and quickly. Such particles do not travel very far into the lungs. Smaller particles below 2.5 micrometres in diameter can be just the right size to become lodged in lung cavities.

Attention has been increasingly focused on these very fine particles (PM25). These can remain in the air for days or weeks and can penetrate deep into the lung, collecting in the alveoli, tiny air sacs where oxygen enters the bloodstream. A number of potentially harmful substances have been found in PM25. For example, elemental carbon can pick up cancer-causing chemicals like benzo(a)pyrene and transfer them to the lungs. An American study suggests an increased mortality risk of 10 per cent for babies exposed to high concentrations of airborne particles (Woodruff, 1997). In 1995, the British Government stated, 'It is clear . . . that the potential effects of particles on human health must be taken seriously' (Departments of Environment, Health and Transport, 1995, p. 2). Following the release of a draft report by the Airborne Particles Expert Group in the summer of 1998, it seemed likely that a PM25 standard would be added to the National Air Quality Strategy.

For all the understandable concern about the link between air pollution and health, considerable media attention was given to a study which found that Saturdays on the east coast of the United States had 22 per cent more rain than Mondays and suggested a possible link with air pollution (Cerveny and Balling, 1998). Because pollutants build up during the week, airborne particles could 'seed' raincloud formation, leading to precipitation. There is no evidence that this effect extends to Britain, although if it did, it might have more impact on public opinion than evidence about health risks.

Environmental challenges for European air quality policy

The European Union's efforts in the area of air pollution were slower to get off the ground than in relation to water pollution. In the 1980s, the emphasis was on the acidification debate and the consequent damage to forests, leading to an emphasis on stationary sources and the development of the Large Combustion Plants Directive. The EC did pass a directive on the approximation of motor vehicle emissions in 1970, followed by another in 1982 and a directive placing limits on nitrogen oxide emissions from motor vehicles in 1977. However, these individual measures fell far short of a comprehensive strategy to deal with air pollution from mobile sources. It was not until 1996 that a framework Directive dealing with ambient air quality was passed and it was 1999 before the first 'daughter' directive passed its final hurdle and was published in the *Official Journal*.

Given the substantial volume of work in the area of water pollution discussed in the previous chapter, why did it take so long to make progress on the arguably equally pressing area of air pollution? The attention given to water pollution and waste disposal issues was in part because the effects of these forms of pollution were more tangible and it was possible to demonstrate clear benefits that could be publicised to citizens (as in the case of cleaner bathing beaches).

There were also important differences in the underlying structures of political power. In the water sector, the Community faced a set of public utilities who were not particularly well organised at a European level. 'The Urban Waste Water Treatment Directive [1991] demonstrated that the waste water sector was not sufficiently co-ordinated across Europe to present a united view at Community level' (Thairs, 1998, p. 161). In any case, because of their monopoly status, they could pass on any additional costs to their users (much the same still applied after privatisation in Britain).

In relation to air pollution, the Commission has had to face the politically powerful oil and motor industries, or industrial interests controlling large plants that contributed significantly to air pollution. These interests would then be defended in the Council of Ministers in a process in which 'each Member State defended its economic rather than environmental interests' (Ikwue and Skea, 1996, p. 85), a process illustrated by the example of the important Large Combustion Plant Directive (LCPD) of 1988 which sought to reduce emissions of sulphur dioxide and nitrogen oxides from power stations, refineries and large industrial plants.

The Large Combustion Plant Directive

Acid rain was the leading environmental issue in the early 1980s. There was widespread concern about damage to trees, but also about impacts on wildlife, buildings and visibility. With opinion in Germany on the subject shifting as the Greens became more influential, a 1982 German law was taken as the basis for a Commission draft directive issued in 1983. The LCPD took five years to enact because of opposition from affected interests represented by the UK Government, although interests in other countries sheltered behind British opposition. Indeed, 'But for electricity privatization in Britain, the measure would, perhaps, never have been agreed' (Ikwue and Skea, 1996, p. 93).

The LCPD was a long-delayed response to concern about acid rain which had been raised as an issue in the 1970s by the Scandinavian status, but achieved greater attention in the early 1980s as concern grew about the death of forests in Germany. Domestic regulations were introduced which disadvantaged the German electricity industry and intensive electricity users. 'Having lost out in the German regulatory process, industry saw EU environmental policy as a possible means of relieving the competitive disadvantages caused by domestic legislation' (Ikwue and Skea, 1996, p. 87).

British opposition was due to the fact that the country was then highly reliant on coal for power generation. The Commission was responsible for early strategy on the directive and sought

> to isolate the UK from its less vocal supporters – essentially the 'cohesion' countries, Greece and Ireland – by offering them significant concessions. In view of the UK's firmness, this strategy failed and became untenable when Spain and Portugal joined the EU in 1986. (Ikwue and Skea, 1996, p. 89)

When strategy passed to the Council presidencies, a more political approach was adopted which emphasised what was feasible rather than what might be evironmentally desirable:

> Instead of a single, uniform SO_2 emission reduction for each Member State by 1995, the presidencies moved towards non-uniform emission reductions, based essentially on political acceptability, to be implemented in multiple stages by the early years of next century. (Ikwue and Skea, 1996, p. 89)

It was the German Presidency which eventually achieved agreement. 'In the end the UK acceded partly because of the need to clarify the financial commitments of the electricity industry prior to its privatisation' (Skea and Smith, 1998, pp. 269–70).

One unintended consequence of the LCPD was to encourage the 'dash for gas' among privatised UK power generators. Gas turbine plants presented 'the established generators with a cost-effective way of meeting the obligations for reducing SO_2 and NOx emissions under the [LCPD], avoiding the expensive retro-fitting of coal-fired plants with flue-gas desulphurisation units' (Collier, 1998c, p. 101). More generally, the LCPD directive failed to achieve the original objective of equalising competitive conditions between the member states because of the subsequent national-level regulatory process. Member states are very protective of their national interests in energy matters and 'a regulatory process which ostensibly covered all countries has resulted in entirely different compliance strategies in the different Member States' (Ikwue and Skea, 1996, p. 94). As was emphasised in Chapter 3, passing the directive is only part of what has to be done if policy initiatives are to have the desired effect on outcomes.

Sulphur dioxide emissions have been cut drastically in both Europe and the US. According to EEA data, they halved in 44 countries between 1980 and 1995. However, the other main component of acid rain, nitrogen dioxide, is much harder to achieve because of growing volumes of road traffic. Acidification remains a major problem, but the attention of environmental campaigners has shifted elsewhere to other issues such as global warming.

The political difficulties encountered by mobile source policies

In some ways 'stationary' sources such as power plants are politically easier to deal with than 'mobile' sources: not only are there a lot more of the latter, but cars are driven by citizens rather than owned by large companies which may be sensitive about their environmental image. The formulation of appropriate policies is made more difficult by the existence of some contradictions in the attitude and behaviour of citizens. For example, research for the Department of the Environment and Transport shows that the number of people 'very worried' about urban air pollution has risen from 23 per cent to 48 per cent in the last ten years (*Air Quality Management*, March 1998, p. 3). Such concerns, however, have to face a situation in which the conditions of use for most individuals are relatively fixed in the short to medium term: alternatives to cars are not available for many journeys or, if they are, they are less convenient or more expensive. Hence, in Britain national road traffic forecasts show that the volume of traffic could increase by up to 84 per cent between 1997 and 2031 unless steps are taken to curb growth.

There are not many effective policy options available to cope with such a massive growth in road traffic. Technical solutions such as cleaner fuels and cars are favoured by industry. However,

> The volume of road traffic in Western Europe is expected almost to double between 1990 and 2010, leading the Dobris Assessment to conclude that the benefits of technological improvements could be canceled out by growth in the number of vehicles on European roads. (McCormick, 1998, p. 196)

Providing new or improved public transport facilities may solve specific problems in particular localities, but not the overall problem. Road pricing has been much advocated by economists, but might encounter political resistance if it was introduced on any scale (which it would have to be to have any impact on road use, and, in any case, public transport systems could not cope with many additional users).

Above all, what may appear to be sensible and feasible policy solutions have to cope with the cultural fascination with the car. A study of young people in Britain by the Chartered Institute of Transport found that most people would rather have a driving licence than the vote if they had to choose, with 76 per cent opting for a licence (Solomon, 1998, p. 6). Sixty per cent of those interviewed saw driving as the most important symbol of coming of age.

The author of the study commented:

> The ability to use a car is not just a means of transport for the coming generation – it is a fundamental area of life. They would strongly oppose any constraints on the use of cars, unless they could see the obvious benefits for themselves from such measures. (*Financial Times*, 14 June 1998)

Technical policy problems

Devising effective air quality management policies poses a number of technical problems. In some areas progress has been made, as with effective measurement techniques, although one consequence has been to expose the seriousness of the fine particle problem. In other areas, technical constraints may not be so readily overcome. In particular, actions taken in individual cities to overcome local problems may simply displace the problem elsewhere or may be ineffective because pollutants are 'imported' from areas a long way away.

It is well known that 'plumes' of ozone precursors may drift considerable distances from urban into rural areas. In urban areas, nitric oxide from vehicles removes ozone from the air. Indeed, modelling work carried out at the Karlsruhe Research Centre in Germany suggests that cutting back on emissions of ozone precursors may increase atmospheric ozone. If, hypothetically, one eliminated local emissions of nitric oxide, the sink for ozone is removed, leading to higher concentrations (*Air Quality Management*, October 1997, p. 8).

It is well known that ozone is no respecter of national boundaries and Europe in particular 'is very susceptible to trans-boundary air pollution phenomena' (UK Photochemical Oxidants Review Group, 1993, p. 111). In anticyclonic conditions, 'vast qualities of precursor emissions from many urban-industrial areas often become trapped above a low-level temperature inversion and drift a thousand kilometres or more across state and national boundaries' (Elsom, 1996, p. 60). In Britain, 'This air often comes from continental Europe and the episodes are more frequent in southern England' (United Kingdom Photochemical Oxidants Review Group, 1993, p. 1).

Recent evidence suggests that concentrations of fine particles in London are strongly influenced by pollution sources in continental Europe. Such concentrations rise sharply in London and Kent when the wind is blowing from the east (*Air Quality Management*, October 1997, p. 5). Other research has also found that periods of high PM10 concentrations are associated with air masses arriving in Britain from mainland Europe (*Air Quality Management*, July 1997, p. 1). If one is dealing with a national and trans-boundary phenomenon, local controls may only deliver a small improvement in air quality. Yet the National Air Quality Strategy in Britain places the responsibility for managing air quality on local authorities. A further problem when looking at particles on a Europe-wide basis is that 'the concentration of airborne particles, PM10, is generally higher in the Mediterranean south than in the cooler northern states. Is it feasible to impose a single PM10 standard on all member states alike?' (*Air Health Strategy*, May 1997, p. 9).

The spread of vehicles with catalysts throughout the car fleet has had a marked beneficial effect on air quality, although catalysers are not effective until they have warmed up and are therefore of little use on the short trips which account for a high proportion of all trips by car. There is no equivalent technological breakthrough available in the medium term. Electric cars are inhibited from becoming commercially attractive by the absence of a satisfactory battery technology, and, as a consequence offer inferior performance for a higher unit cost. Lead acid batteries offer a high weight-to-power ratio, and although there are a number of alternatives available, they are currently too expensive for commercial use. Increased electricity production to meet the needs of electric cars would also have implications for global warming.

Even incremental improvements in existing technology pose some difficult problems:

> The relationships found among fuel properties/engine technologies/exhaust emissions are complex. Changes in a given fuel property may lower the emissions of one pollutant but may increase those of another (i.e., decreasing aromatics content in gasolines lowers CO and HC emissions but increases NOx emissions). In some cases, engines in different vehicle categories, such as heavy duty and light duty vehicles, have disparate responses to changes in fuel properties. (ACEA/Europia, 1995, p. 3)

One short-term measure might be to tackle the dirtiest vehicles through roadside checks and scrapping subsidies. It has been estimated in Britain that 10 per cent of vehicles are responsible for 55 per cent of emissions (*Air Quality Management*, January 1998, p. 4).

In a pilot fixed penalty programme in Britain, 6 per cent of 15 000 vehicles stopped were outside prescribed limits for tailpipe emissions, with the worst offenders being diesel cars, vans and taxis. One-third of the taxis stopped in the London borough of Westminster could not be examined because they were unfit to take the test (*Air Quality Management*, January 1999, pp. 8–9).

Such programmes face enforcement constraints. They require the cooperation of the police, an often neglected actor in analyses of air quality management policy (Marek, 1999) who may have other priorities. An extended pilot programme in Britain led to an adverse public reaction because motorists were not offered the opportunity to rectify any faults. Replacing the fixed penalty by a rectification option would reduce the revenue flow and undermine the precarious funding of the scheme. Motoring organisations argued that until on-board diagnostic systems are available to inform drivers that their vehicle is exceeding agreed limits, it is unfair to fine them at roadside emissions tests. Such a policy approach also encourages motorists to take the view that air pollution is not 'their' problem, but is caused by a minority of old or badly maintained vehicles.

The evolution of EU policy on air quality

The development of the European Union's policy on air quality has not occurred with any great urgency and until the measures taken in the late 1990s, the policy was clearly inadequate:

> Unravelling the EU's tangled policy on vehicle pollution is as hard as cleaning a sooty engine. After trailing the US and Japan, the European Commission issues its first directive on 'regulated substances' in exhaust emissions in 1970. It has been catching up ever since. (*Financial Times*, 20 August 1998)

Throughout the 1980s and early 1990s the Commission passed a series of directives which set 'guideline' and 'limit' values for five pollutants: nitrogen dioxide, sulphur dioxide, particulates, lead and ozone (the directive on this important pollutant was not passed until 1992). 'Limit Values or Standards are legally binding levels which should not be exceeded in any area of its jurisdiction' (European Environment Agency, 1997, p. 57). The reality was, however, somewhat different than this bald statement implies.

In practice EU law often

> imposed no real costs because sufficiently stringent standards were already in place under national legislation ... And in many cases national pollution levels were far below those mandated by EC law, despite an absence of national legislation. (Golub, 1997, p. 6)

For example, take the first directive in the series, that on sulphur dioxide:

> standards already met by Belgium, Denmark, Germany, Netherlands ... France incurred no sweeping changes or costs ... Italy's emissions were already falling ... the UK had no SO2 problem. (Golub, 1997, p. 7)

Even if member states were in breach of the standards, little action followed. 'Technically speaking, member states are required to report to Brussels any breach of the limit values. Traditionally, however, that has tended to be the end of the story' (*Air Health Strategy*, May 1997, p. 8).

The Commission recognised that this approach of tackling one pollutant at a time, and taking very little in the way of follow-up action, had failed. The Five Directives 'differ in the times when they were adopted and the philosophical basis on which they were agreed and do not provide an overall view of the situation regarding ambient air quality in the European Community' (European Commission, 1994, p. 11). Decoded, this means that the directives lacked coherence and there was no overall policy.

The underlying political problem was that 'long term air quality objectives have not been considered to be very important by the majority of Member States' (European Commission, 1994, p. 2). The Commission recognised that the implementation of the existing directives had revealed a number of problems. These included the fact that compliance was not achieved as quickly as possible and the 'large differences in monitoring strategies in comparable situations between and within Member States' (European Commission, 1994, p. 2). The Commission therefore proposed a new strategy based on new EU-wide air quality objectives, as well as setting minimum standards for monitoring air quality and disseminating the results to citizens.

This led to the Air Quality Framework Directive of September 1996. This established new EU-wide air quality objectives, as well as setting minimum standards for monitoring air quality and passing the results on to the public. The framework Directive requires 'long-term limit values' (LTLVs) for air pollution to be met within 15 to 20 years with provision for interim 'current permitted values' (CPVs) as well. Areas will be classified by member states into three types. In areas of poor air quality where both sets of values are exceeded, CPV must be achieved as soon as possible and LTLV within a specified time period. In areas of improving air quality where pollution is between the CPV and the LTLV, the LTLV must be achieved within a specified period. It is only in the areas of good quality where air pollution is below the LTLV that no action is required:

> A crucial difference between the previous regime and the new air quality arrangements is that in future, member states will be required to draw up plans for protecting the public from polluting episodes, rather than merely reporting any exceedences to Brussels. (*Air Health Strategy*, May 1997, p. 8)

The task of drawing up 'daughter' Directives for each of 12 pollutants was assigned to working groups made up of representatives from industry, member states and non-governmental organisations. The 'daughter' Directives will cover not only limit values and target dates, but also give guidance on monitoring and assessment procedures. The first four draft directives, covering SO_2, lead, particles and NO_2 were made public in the autumn of 1997. The limit values set were substantially based on World Health Organisation standards. One of the features of the proposals was a 'margin of tolerance' which decreases in linear fashion as the target date approaches.

The first four 'daughter' Directives setting objectives and limit values for sulphur dioxide, NOx, particulates and lead were approved by the Council of Ministers in June 1998 and were finally adopted after a prolonged passage through the EP in June 1999. Member states then had two years to enact the proposals. They provide for target dates in January 2005 or January 2010 in the case of nitrogen dioxide. The Council accepted an amendment from the Parliament requiring an alert threshold for nitrogen dioxide which will require member states to issue warnings if it is exceeded. However, the Council voted to loosen the terms of the hourly limit value for nitrogen dioxide so that it can be exceeded 18 times a year rather than eight as originally proposed. This brings it closer to the provisions for sulphur dioxide where the limit value can be exceeded 24 times a year and particulate matter where the limit value can be exceeded 35 times a calendar year. The Council also rejected the Parliament's suggestion of an alert threshold for PM10 'arguing that no "safe" limit had been shown to exist for this pollutant' (*Air Quality Management*, July 1998, p. 3). This might be seen as sidestepping one of the most serious problems. The next two 'daughter' Directives will cover carbon monoxide and benzene and will be dealt with in the EP through the co-decision rather than the cooperation procedure.

Much has been achieved in legislative terms, but it will some time before there is any considerable impact on outcomes, 2010 for most pollutants. Even then, it will be possible for 'exceptional' episodes of pollution to occur without breaching the guidelines. The EU is also likely to face the dilemma of what action to take in 2010 when it is more than likely that it will be found that a number of areas in the Community exceed the permitted values.

AutoOil

The AutoOil programme instigated by the European Commission in 1993 was hailed as a new, non-confrontational approach to obtaining the drastic cuts in emissions required to meet World Health Organisation standards. It thus reflected the move away from a 'command and control' regulatory style to the use of NEPIs referred to in Chapter 1. The discussions were confined to the Commission and the oil and motor industries, excluding the non-governmental organisations that had usually been incorporated in the policy-making process.

The idea behind AutoOil was that the European Commission, the motor industry and the oil industry would work in a partnership to tackle road transport's contribution to air pollution through cleaner fuels and curbing vehicle emissions:

> By working together in this way, EU officials hoped to avoid the political infighting and furious industry lobbying that occurred when new fuel and vehicle specifications were introduced in the US under the 1990 Clean Air Act. (*Financial Times*, 26 June 1996)

The objective of the programme was stated as being 'to identify which new measures may be required to meet rational air quality objectives in the most cost effective way, derived from scientifically sound data' (ACEA/ Europia, 1995, p. 2).

One of the consequences of this partnership was the European Programme on Emissions, Fuels and Engine Technologies (EPEFE) to which the car and oil industries contributed ECU10m:

> The EPEFE programme was designed to extend the information on the relationships between fuel properties and engine technologies and to quantify the reduction in road traffic emissions that can be achieved by combining advanced fuels with the vehicle/engine technologies under development for the year 2000. (ACEA/Europia, 1995, p. 2).

This statement is significant for what it omits. Unlike the air quality programme in California which has had a (possibly mistaken in the medium term) emphasis on developing alternative vehicle technologies (Grant, 1995, 1998), this programme is clearly based on the further development of existing technologies. In other words, the emphasis is on improving the environmental performance of the gasoline-fuelled engine, meaning that motor manufacturers and oil companies will not have to sacrifice their sunk costs in existing technology.

Hopes for a speedy agreement on appropriate measures were dashed when it was revealed as the Commission's proposals were announced in 1996 that cutting emissions by 70 per cent would cost the motor industry ECU4.1bn a year over 15 years and the oil industry just ECU766m. Those involved in the process thought that one reason that the motor industry fared badly in the early stages of the decision-making process was poor lobbying. According to one environmental civil servant, 'Oil has an advantage because it only really has one product whereas the motor manufacturers are competing with each other head on and have more internal divisions. But the oil industry was more astute' (*Financial Times*, 17 February 1998).

The motor industry was beset by recurrent problems of overcapacity in Europe throughout much of the 1980s and 1990s and was not going to assent willingly to any changes in its products that would push up costs that might not be readily recouped on a saturated market. As a consequence, there was a two year delay between the announcement of the AutoOil proposals on tailpipe emissions and fuel quality standards for new cars and their final adoption in 1998, making the whole process as long as that for the LCPD. This delay occurred because the package was 'the subject of two years of wrangling between the motor industry and oil producers, with each side claiming the other should shoulder more responsibility for reducing emissions from new cars' (*Air Quality Management*, July 1998, p. 5). The motor industry became more effective in the second phase of lobbying, with Volkswagen, Daimler-Benz and BMW exerting their influence with German MEPs putting the argument that further substantial cuts in pollution could only come with cleaner fuels. In particular, MEPs became convinced by the motor industry's argument that 'oil companies need to implement drastic cuts in fuel sulphur content to enable the efficient use of "denox" technology for cutting nitrogen oxide' (*Financial Times*, 17 February 1998).

The final cost is likely to work out at 60bn euros for the motor industry and 32bn euros for the oil industry over the 15-year implementation period. The European Petroleum Industry Association commented, 'The final agreement has moved significantly from the Commission's proposal of June 1996 that Europia supported' (*Financial Times*, 20 August 1998). According to Europia, the earlier proposals would have cost its members a third of the bill they now faced.

The objective of the package is to cut emissions of gases such as nitrogen oxides, ozone, benzene and carbon monoxide by around 70 per cent on average. This is to be achieved by a two-step tightening of emission limits for passenger cars and light vans – the first in 2000, the second in 2005. By 2005 the EU will have standards that are broadly similar to California, but with a tougher limit for nitrogen dioxide. Leaded fuels will be phased out by 2000 in most cases. In addition the package will entail the fitting of on-board diagnostic systems to new petrol cars from the year 2000 so that emissions can be checked, a requirement to be extended to diesel cars in 2005. Oil companies will be required to introduce 'reformulated' fuels by early next country. By 2005 there should be a threefold reduction in the sulphur content of petrol and a seven fold cut for diesel.

Under pressure from the Parliament, the final package agreed in September 1998 was tightened up, although some observers saw the concessions as marginal (Peterson and Bomberg, 1999, p. 179). Indicative limits for exhaust emissions in 2005 were made mandatory and the indicative target for sulphur content in fuel will also become mandatory, although at the same level as the indicative targets. The Commission had wanted to await the outcome of a second autooil programme before setting mandatory standards for 2005. However, the conciliation process meant that Parliamentary representatives had to back down on many of their original demands. The overall plan is an ambitious one and represents one of the most significant steps forward to date in the improvement of European air quality.

Impact at the member state level

In many ways the most important impact of the EU's directives on air pollution at member state level has been in terms of the way that they have not just initiated reactive changes in policy content, but also changed the way in which policy is made. In response to EU initiatives, policy networks have changed their shape and reformed, setting in train dynamic processes which may have even more far-reaching effects in the long run. An interesting contrast in this respect may be drawn between France and Italy. France is a highly centralised country with a well-established, if somewhat rigid, bureaucratic apparatus. Italy is much more decentralised, particularly in terms of the real distribution of authority, and suffers from a number of problems in terms of the effectiveness of its bureaucracy which can make it difficult to implement new policies.

In the case of France, the country had a traditional and closed air quality policy network which was concerned with problems of pollution from industrial sources occurring mainly in the winter months. However, by the end of the 1980s, both the clean air policy programme and the composition of the policy network had changed 'mainly under pressure from the EEC, but also, to a lesser extent, under environmentalist and social pressure' (Larrue and Vlassopoulou, 1999, p. 102). The 1991 decree which integrated four of the EC air quality directives into French law 'constituted another step toward the opening of the air policy network to car pollution problems' (Larrue and Vlassopoulou, 1999, p. 103). The subsequent ozone directive played a major role in increasing public awareness of air pollution problems (Larrue and Vlassopoulou, 1999, p. 106). Larrue and Vlassopoulou conclude (1999, p. 105) that the recent history of air pollution policy in France

> can be characterized by an external change linked to the intervention of an external actor, the EEC. As a matter of fact, . . . the change of problem perception among air policy as well as transport policy networks, has been initiated by EEC directives. As a consequence, this supranational intervention has modified the configuration of the air pollution policy network in France.

In Italy, the policy style has been highly reactive and bureaucratic inertia has been a significant constraint in implementing effective policies. There have been considerable delays in spending funds allocated for mass transport systems and even bicycle paths which is explained in large part by 'The poor performance of public bureaucracies and cumbersome administrative procedures' (Desideri and Lewanski, 1999, p. 68). Legal constraints have inhibited policies regarded as standard policy instruments elsewhere – for example, until recently it was illegal to charge for parking where there was no attendant. 'Italian governments have limited themselves to enacting, often with considerable delay, EU legislation concerning fuel characteristics and vehicle emission standards' (Desideri and Lewanski, 1999, p. 70). The state legislation 'that local authorities have to refer to is often confused and contradictory' (Desideri and Lewanski, 1999, p. 70). Policy has often been made on an emergency basis in relation to particular crises. However, one positive result has been an increase in a previously low level of public awareness about air pollution problems:

> The fact that traffic causes dangerous levels of pollution that constitute a menace to public health, once an argument shared only by a few scientists and the environmentalists, is now accepted as a matter of fact by policy actors and the population in general. Furthermore, traffic has become a hot topic of political debate, as shown by the . . . electoral campaigns in many cities. (Desideri and Lewanski, 1999, p. 71)

In very different national settings, changes in EU policy have thus had positive effects, not just in policy content, but also in policy awareness and formation (Britain anticipated the policy changes by introducing a new National Air Quality Strategy). Italy is seriously constrained by the lack of a professional technical bureaucracy in the air quality pollution field and policy networks have not changed and stabilised to the extent that has occurred in France. On the other hand, the less centralised policy style in Italy allows more scope for local policy experiments. In both countries, however, EU directives were a significant part of a process which led to greater public awareness of air pollution problems. This in turn led to greater public pressure for policy change with systematic use being made of the referendum at local level in Italy (Bobbio and Zeppetella, 1999, p. 74). EU policy developments may thus set in train political processes which create renewed public pressure for effective action on air pollution.

The possibilities of a virtuous cycle are, however, undermined by the simultaneous public desire to be able to experience the benefits of the motor car without any of the disbenefits such as air pollution. The underlying problem has deep cultural roots:

> if getting into a car and going somewhere is a cultural norm rather than a transport choice, much more fundamental measures will need to be taken, if behaviour is to be changed, than most of those currently being implemented . . . We face a problem that is fundamentally different from persuading people to change their travel modes, and that is persuading them to change their current methods of filling their time and also their cultural symbols. (Solomon, 1998, p. 8)

However, the suggestion that we should 'substantially reduce the comfort and convenience of private motorised travel' (Solomon, 1998, p. 8) is unlikely to be politically palatable. As a stylised fact, technological solutions do just enough to keep pace with the growth in miles travelled by car. The EC thus faces an acute dilemma: the citizens it wishes to protect from air pollution are the principal causes of that air pollution. More targeted and subtle policies which took account of social differentiation in the population might be more effective (EPSECC, 1998). There is no escaping the fact, however, that ground-level air pollution is the area of environmental policy which faces the EC with some of its most difficult challenges in terms of devising acceptable, feasible and effective environmental policies.

Conclusion

In the Introduction to this volume we set out eight criteria which might be used to judge whether a policy is effective. What does our review of European environmental policy tell us about how well these criteria have been met?

The first criterion related to the clear establishment of authority at a European level. In the EU, where environmental policy competence is divided between supranational and member state authorities, the setting of policy parameters for environmental protection has been bound up in a debate over the limits of what is provided for in the Treaty. Environmental policy now has a clear treaty basis and the environment Directorate has been able to take a proactive role, overcoming some of the initial suspicions about its 'green' orientation.

Problems remain, however, in terms of the range of policy instruments available to the EU and some restrictions in its range of competence. As is evident from a number of chapters in this book, particularly that on climate change policies, member states have made increasing use of the subsidiarity principle to attempt to renationalise environmental policy. EU environmental policy has been distinguished from national environmental policy by the limited range of instruments traditionally used to achieve standards or specific environmental policy objectives. EU environmental policy was traditionally heavily reliant on command and control mechanisms in the form of legally binding standards prohibiting or directing activities within the member states. These often failed to achieve their objectives and incurred substantial transaction costs. The Fifth Action Programme contained a commitment to broaden the range of regulatory instruments (European Commission, 1993, p. 5) and attempts have been made to use a range of NEPIs, for example based on the use of economic instruments (Bell, 1997, pp. 117–24; Scott, 1998, pp. 44–63; Andersen, 1994).

One important instrument, taxation, is not available to the European Union. Taxation is not without its problems as a policy instrument. Energy taxes which are designed to curb consumption by increasing costs are unpopular with the electorate in general and with industry in particular. The introduction of VAT on domestic fuel in the UK proved to be a sufficiently politically sensitive issue for the incoming Labour administration in 1997 to immediately reduce the level of taxation on the grounds that it amounted to a regressive tax which disproportionately affected the old and less well-off in society. The annual 'escalator' increase in road fuel tax is also encountering increasing political resistance. Nevertheless, these difficulties should not disguise the fact that the EU currently lacks the ability to use a significant policy instrument in the range of options available to it, the ability to impose environmental taxes.

The EU's policy authority is also limited in terms of scope. Some key issues in the environmental sphere are virtually absent at EU level. An examination of the successive framework programmes reveals that the sectoral coverage of environmental policy includes environmental assessment, integrated pollution control, atmospheric pollution, waste management and nature conservation. Yet absent from the range of EU environmental policy instruments is an attempt to address land use control (or town and country planning as it is known in the UK) in any meaningful way. Despite the fact that land-use planning is of crucial importance to environmental policy objectives in relation to the built environment and protection of the countryside, it remains within the policy domain of the member states rather than being within EU competence.

If the subsidiarity principle is applied correctly, it is perhaps only right that development plans, building regulations and applications for planning permission should be dealt with via a combination of national and subnational policy competences.

However, there are a range of business actors who complain that a level playing field will never operate effectively in the single market until the EU takes greater responsibility for removing some of the planning anomalies that presently amount to non-tariff barriers to trade in the EU. In the retail sector, for example, changes in government policy relating to the building of new out-of-town shopping stores are claimed to be a hindrance to penetration of the domestic market by non-UK food retailers (Mayes and Hart 1994), while in the construction sector UK contractors have complained that building regulations, particularly in southern Europe, have been applied more rigorously against them than against domestic business firms (Matthews and Pickering, 1995). Conversely, national or regional authorities which apply land-use planning regulations in a less than rigorous manner in relation to indigenous firms than in relation to new market entrants may encourage unnecessary environmental degradation because priority is given to concerns about commercial competitiveness and national economic well-being.

More generally, the retention of land-use planning regulations as national and subnational policy competences is further evidence of the tension between market integration and the greening of the EU, an issue which we return to later in the conclusion. While the anomaly of land-use planning regulations as an exclusive national and subnational policy competence remains, the environmental policy authority of the EU is undermined by incomplete coverage and a failure to address the full range of issues that arise from an objective of comprehensive environmental protection.

The second criterion calls for a rules-based policy backed by tough European law. No one could doubt that, for example, in the area of water pollution, there is a comprehensive rules-based policy, and an equivalent framework of rules is rapidly being developed in the area of ground-level air pollution. Nevertheless, in the case of water policy, the EU has consistently failed to improve water quality standards in rivers and lakes to such an extent that 25 years of regulatory activity amounts to a case of regulatory failure. This has led to re-regulation in the shape of a new flexible and decentralised framework approach. Perhaps one of the lessons to be learnt from European environmental policy is that too many rules can be counterproductive and too tough a stance can simply encourage resistance. Hence, the emphasis on NEPIs such as partnership agreements with business.

The third criterion is that policy should be backed by resources which are distributed at the European level providing incentive structures. EU environmental policy is essentially regulatory so any resources it distributes are created indirectly. For example, the development of an environmental technology industry or a more sophisticated recycling industry has undoubtedly been encouraged by EU regulations. However, there is no clearly identifiable client group in the way that farmers may depend on EU subsidies to stay in business. The benefits to citizens may also not be immediate or tangible when compared with an area like social policy where directives on working hours or parental leave can have a real and discernible impact on the everyday life of individuals.

The absence of a clearly defined clientele group does make it harder to maintain the momentum of the policy. This is exacerbated by the fact that environmental policy has lost some of its dynamic. This is in part because the focus of attention has switched to economic and monetary union and the enlargement process and in part because the failure to proceed with a carbon tax was a watershed decision which encouraged an emphasis on incremental progress in specific policy areas rather than 'great leaps forward'. As the analyses of individual policy areas have shown, significant initiatives are taking place in water policy and air quality policy. The momentum is, however, very much a bureaucratic one: a framework Directive is passed and this in turn is followed by a series of 'daughter' Directives which eventually emerge after a long process of consultation with affected interests.

The latest batch of countries to join has not had the environmentally progressive impact than had been hoped for and, even with an environment minister from the Green Party, Germany has to balance environmental policy against economic concerns such as the impact of decisions on its motor manufacturers. The continuing presence of the Greens in the EP continues to ensure that it is a relatively environmentally proactive body, but its influence in the decision-making process, although growing, is still limited (not least in relation to the environmentally damaging CAP). The 'green' pressure groups have a media presence, some capacity for mass mobilisation and knowledgeable staffs, but their influence is offset by that of business interests much of the time. The public is giving environmental questions less priority than ten years ago. In many ways, the whole policy area has been routinised, bogged down in the production of new Directives, or in interminable international negotiations about global issues such as climate change which have an uncertain outcome.

The fourth criterion is a crucial one, that policy actually changes the behaviour of relevant actors. It is here that we encounter the implementation and enforcement problems that beset environmental policy which we reviewed in Chapter 3 of this book. Even if member state governments and national agencies are sincere about wanting to implement policies properly, there is often a big gap between intention and enforcement. We noted the risks of 'free riding' problems in EU environmental policy if actors are able to enjoy the benefits of environmental improvements without themselves complying with the new standards. We saw how the intended policy implications of environmental initiatives may be hindered not only by selective compliance strategies on the part of business, but also by the failure of member states to implement and enforce environmental regulations effectively. In relation to environmental duties, where member state responsibilities are often delegated to subnational agencies with discretionary powers, we found some evidence of regulatory failure that supports Weiler's view (1991) that poor enforcement is the 'black hole' in the EU policy process.

Given the limitations of what can be done by the Commission and the ECJ to improve member state implementation and enforcement through the formal procedures set out in Article 226 (formerly Article 169) of the Treaty, we have suggested that resource problems might best be addressed by improving access to environmental information, with better transparency of environmental data through the activities of the EEA. We have also suggested that improving the effectiveness of EU environmental policy could be achieved by improved coordination of best practice and information exchange between environmental enforcement agencies through EU-wide initiatives such as IMPEL.

The fifth criterion advanced was that policy needs to be based on innovative ideas which produce remedies that enjoy intellectual integrity. The nature and degree of action required by the ecological imperatives of a problem will be filtered both by the framing processes of the experts who identify and package the problem for policy resolution and the policy community which is expected to take action on the problem. The community will inevitably interpret the problem within the context of existing institutional mindsets shaped by previous experience of dealing with similar environmental problems, the interests of the policy-makers themselves and their perceptions of what concerns the constituencies to which they are accountable.

The point is that no amount of scientific knowledge or data on a problem will improve the ability of the EU to respond to environmental problems. According to this reading, optimism about the role of the EEA in advancing EU environmental policy may be somewhat misplaced. Policy responses will be strongly affected by institutional biases, the competing frames of different actors (Jachtenfuchs, 1996) and the interests of the knowledge community themselves. The mobilisation of bias and the practice of non-decision-making will also ensure that some issues never enter the policy-making process.

Environmental problems are more than technical issues that can simply be addressed through science and technology. They are first and foremost political because they affect social groups differentially and impose different types of cost and burden. Strengthening the knowledge base which informs policy will not of itself, in the absence of heightened political will, lead to more effective policy. What is deemed to be ecologically effective is contested by those who provide the knowledge of environmental problems and those expected to respond to them.

The sixth test we borrowed from Wallace was an implicit Pareto optimum, representing an equilibrium point where everyone relevant is as well off as possible. This is difficult to operationalise, but much of the discussion in the book suggests that outcomes are often suboptimal when viewed from the perspective of our knowledge about what is necessary to protect the environment. In part, this because there is not a single environmental policy which seeks to protect the ecosystem as a balanced whole, but rather a series of policies which tackle specific problems. To some extent this is unavoidable because the problems posed are distinct and require rather different solutions. However, one might question whether the policy has become too disaggregated and whether the sum of its parts add up to less than a whole. For example, it was suggested that this was the case with climate change policies.

A case study by case study review of policy shows that the EU has 'greened' at a different pace depending upon the issue in question. That this is so should not be at all surprising given that some environmental problems are less well understood and therefore scientific underpinnings are more readily contestable. Some are also more economically costly in terms of the scale of restructuring they may require, while others are far more amenable to a technical fix. It is expected, for instance, that certain air quality issues may be treatable through available or easy-to-develop technological solutions. Others, such as climate change, are altogether more difficult to address in terms of the changes in human behaviour required and the indictment of energy intensive economic growth which it implies.

It may be, however, that the EU has arrived almost unintentionally at an approximate but rather suboptimal equilibrium position in terms of its attempts to offer its citizens economic success and certain social standards with some environmental protection thrown in. This form of 'squaring the circle' of incompatible objectives would seek to eliminate the grossest forms of pollution by promoting and encouraging forms of technology which reduce environmental impact. The encouragement of practices such as recycling reassures citizens that they have 'done their bit' for the environment as they drive away from the recycling point. Environmental attitudes vary across the EU and are not always deeply held. In a sense, the EU is giving its citizens what they want: enough action on the environment to assuage their consciences, but not so much as to threaten their livelihoods and lifestyles.

Indeed, this discussion of what constitutes a politically acceptable if ecologically suboptimal equilibrium leads us into the seventh criterion which argues that even if the policy is suboptimal, the alternatives may be worse. It is easy through criticising the shortcomings of EU environmental policy to slip into a position which overlooks its achievements. If there was no EU environmental policy, levels of environmental damage and destruction would probably be worse. The EU has helped to make some rivers cleaner, and bathing waters somewhat safer, and it may yet make the air more breathable. If limited progress has been made on global warming, it has not been for want of effort and commitment.

The pessimistic interpretation of EU environmental policy is that it is no more than a paper policy, a series of legislative enactments produced by a legislative factory which are then put into effect in a half-hearted and limited fashion. This interpretation would argue that the 'greening' of Europe is simply an illusion provided by a set of green-tinted glasses which put an EU driven by the interests of multinational companies in a more favourable light. The optimistic interpretation is that the cumulative effect of all the incremental policy changes will be such as to produce a set of conditions where the environment is significantly more protected and people have adopted more environmentally sustainable lifestyles. That day is some way off. There is a process of 'greening' going on, but evidence of its impact is often hard to detect. Whether that is better than the alternatives, or whether it reduces pressure for more radical change, is a judgement that will be informed by alternative perspectives.

The final criterion of effectiveness raises the question of whether the policy serves predominantly symbolic goals or whether it also has a real substantive impact. In order to address this question we need to raise the issue of ecological effectiveness and in particular whether environmental considerations are starting to have an impact in other policy areas (what is known in European jargon as 'the Cardiff process'). This in turn raise the question of the tension between the commitment to economic growth and competitiveness which remains at the heart of the European vision and the pursuit of environmental goals.

'Greening' can take place on a number of levels. For many, the vast body of legislation relating to environmental protection that already exists within the Community provides sufficient testimony to the fact that the EU has 'greened'. There is evidence of a 'light green' transition having taken place away from the more excessively destructive pollution-intensive practices of the past. Effective policy responses in the areas of water pollution and ozone depletion (to name but two) have been formulated and implemented. And internationally, the EU is regarded as a leader: the principal catalyst to action on a range of global environmental threats.

On another level, however, the singular failure to respond to repeated requests, emodied in the Environmental Action Programmes of the EU and now in the treaties, to integrate environmental policy imperatives into all other policy areas, makes it abundantly clear that a genuinely cross-sectoral 'greening' has not taken place. Most areas of policy for which the EU is responsible proceed in a manner largely uninformed and untouched by the need to internalise environmental costs. For example, the creation of Trans-European Road Networks that generate vast quantities of CO_2 undermine the efforts of the environment directorate and others to green the EU. A task force of independent experts appointed by the Commission calculated that the growth stimulus of the internal market would result 'in a 8–9 per cent increase in SO_2 emissions and a 12–14 per cent increase in NO_x emissions by the year 2010' (Klatte, 1999, p. 2).

The policy that still accounts for nearly half of the budget, the CAP, provides incentives for ecologically damaging farming practices. There is a continuing debate about making the policy more environmentally friendly, and this is to some extent reinforced by the possibility that payments linked to farming practices that protect the environment may be permitted under international trade rules. The 'green box' may thus become truly green. New rural development legislation allows member states to choose from a menu of options but makes an agri-environmental element compulsory. Nevertheless, for the time being, the CAP continues to encourage or permit forms of farming that pollute rivers or threaten biodiversity. This outcome is not accidental, but reflects the fact that North European grain farmers can only compete with their North American rivals through chemically intensive forms of farming. There is an economic dynamic there which, for all the talk of ecological modernisation, is fundamentally at odds with environmental goals.

On a broader level still, we have noted the continued privileging of trade and economic objectives above those of environmental protection. Indeed, those initiatives most likely to succeed are those which simultaneously serve economic objectives in terms of creating new markets for European exporters of clean technologies, act as a barrier to competitor imports or which provide employment and therefore 'win–win' opportunities (Golub, 1998). If environmental policy can be shown to have favourable effects on employment, for example, through the creation of industries producing new environmental technologies, its chances of successful adoption may be increased (Barnes, 1999).

Taking this argument a step further, it is possible to question whether the EU, with its origins as an economic bloc designed to lower tariffs between its members, is ever likely to lead the world in environmental affairs. For some ecologists, the goal of economic growth which drives the project of European integration is an unsustainable one. In other words, as long as the EU intends to expand production and consumption of resources year on year, it will never be in a position to address the root causes of environmental degradation, but instead will be satisfied with post hoc clean ups, technological fixes and other end-of-pipe solutions.

It is difficult to see the EU performing a different role in the global economy, given that its very raison d'être is the liberalisation of trade. In other words, the limits of the 'greening' of the EU are a feature of the process it supports and continues to accelerate. From this perspective, the rhetoric of ecological modernisation obscures the causes of problems which arise from the routine and mundane practices of European economic integration.

The nature of this fundamental conflict is visible in a problem already referred to: that the making of policy in areas other than those officially designated 'environmental' has largely been without regard to the ecological implications of such activity – a fact which undermines the overall effectiveness of 'official' environmental policy. This is in spite of numerous pronouncements by the EU calling for the integration of both objectives ranging from as early as the First Action Programme on the Environment (Wilkinson, 1997) to the so-called 'Cardiff process', which has yet to make a substantial impact. The possibility of their meaningful integration rests on a notion of ecological modernisation (Weale, 1992) which assumes that economic development and policies to protect the environment are complementary. Whereas from an ecological perspective, the fact that EU environmental policy has not come very far is understood as a feature of the tension at the heart of the Community which clings to the conflicting ambitions of resource-intensive economic growth *and* effective protection of the environment.

The liberal counter-position to this would be that liberal trade policies promoted by the EU have led to a raising of standards not just within Western Europe, but that globally there has been a 'trading up' of standards rather than a process of competitive deregulation. Writers such as Vogel (1989) point to the extensive body of EU environmental law to illustrate the compatibility between the objectives of trade liberalisation and environmental protection. Whilst this work clearly highlights the extent to which inter-state or state–firm trade can 'ratchet up' standards, it remains difficult to contest that the form of economic development that the EU has promoted has 'cast a long ecological shadow' (Bretherton and Vogler, 1997).

Given the economic structures and goals which drive the EU, it is possible nevertheless to argue that its environmental policy is incredibly successful in the light of these constraints. It is unquestionable that for a body whose founding Treaty does not even mention the environment, the EU has generated an impressive body of environmental law over the past two decades. The challenge then is to enquire into why it is that environmental policy is not more effective given that many of the technologies that are claimed to be adequate for dealing with enviromental problems already exist, in most cases the knowledge base necessary to guide action is sufficiently consolidated and the policy instruments (standards, permit trading, and, to a more limited extent, taxes) to tackle the problems are at the disposal of policy-makers who choose to use them. Explaining inaction and the routinised non-decision-making practices of the EU is as important as accounting for those environmental policies which are in place.

What we have focussed on in this book is what has happened in the areas of policy formally ascribed the term 'environmental' – those which the EU itself would recognise as environmental policies. What our analysis has shown, however, is that many of the forces and factors which militate against more effective environmental policy relate to actors and pressures in non-environmental areas of policy such as trade, agriculture and energy. Until decisions in those areas are also guided by a concern to protect the environment, effective environmental policy will remain an illusion.

The irony, of course, is that in order to do this, the traditional path of integration would have to be challenged. There would need to be a greater emphasis on the costs as well as the benefits of economic growth. How the EU chooses to confront these central dilemmas takes on global significance. The EU's ability to project a credible and sustainable model of development applicable to the aspirations of the majority of the world's people rests on its willingness, as yet undemonstrated, to address many of the contradictions at the heart of its operations which currently frustrate more effective environmental policy.

Bibliography

ABB (1997) *Annual Report 1997: ABB Group and Parent Companies* (Zurich: Asea Brown Boveri).

ACEA/Europia (1995) 'European Programme on Emissions, Fuels and Engine Technologies', European Petroleum Industry Association, Brussels.

Acid News (1994) 'Power plants pilloried', 2, April.

Agence Europe (1998) 'Kinnock anticipates communication on air transport and the environment', 5 May.

Air Health Strategy (various issues) (London: Information for Industry).

Air Quality Management (various issues) (London: Information for Industry).

Andersen M.S. (1994) *Governance by Green Taxes* (Manchester: Manchester University Press).

Andersen, M.S. and L.N. Rasmussen (1998) 'The Making of Environmental Policy in the European Council', *Journal of Common Market Studies*, 36, 585–9.

Anderson, D. (1988) 'Inadequate Implementation of EEC Directives: A Roadblock on the Way to 1992?', *Boston College International and Comparative Law Review*, 11, 91–6.

Anderson, D. (1995) 'Rapporteur's report of workshop presentations and discussions' in M. Grubb and D. Anderson (eds), *The Emerging International Regime for Climate Change* (London: RIIA).

Aspinall, M. and Greenwood, J. (1998) 'Conceptualising Collective Action in the European Union: an introduction' in J. Greenwood and M. Aspinwall (eds), *Collective Action in the European Union* (London: Routledge), 1–30.

Audley, J. (1997) *Green Politics and Global Trade: NAFTA and the Future of Environmental Politics* (Washington DC: Georgetown University Press).

Axelrod, R. (1990) *The Evolution of Cooperation* (London: Penguin Books).

Bach, W. (1995) 'Coal Policy and Climate Protection: Can the Tough CO_2 Reduction Target be met by 2005?', *Energy Policy*, 23, 1.

Bachrach, P. and Baratz, M. (1962) 'Two faces of power', *American Political Science Review*', 56, 947–52.

Baker, S. (1996) 'The scope for east–west cooperation' in A. Blowers and P. Glasbergen (eds), *Environmental Policy in an International Context: Prospects* (London: Open University Press).

Barnes, P.M. (1999) 'The Treaty Versus the "Ideal World" – Employment and the Environment', paper for the Sixth ECSA Biennial International Conference, June, Pittsburgh, Pa.

Barrett, C. (1996) *Sustainable Development Case Study Teaching Politics* (York: York University).

Baumgartl, B. (1997) *Transition and Sustainability: Actors and Interests in Eastern European Policies* (London: Kluwer Law International).

Bell, S. (1997) *Ball and Bell on Environmental Law* (London: Blackstone Press).
Benedick, R.E. (1991) *Ozone Diplomacy: New Directions in Safeguarding the Planet* (Cambridge MA: Harvard University Press).
Bergesen, Helge et al. (1994) *Implementing the European CO_2 Commitment: A Joint Policy Proposal* (London: RIIA).
Beuermann, C. and Jaeger, J. (1996) 'Climate Change Politics in Germany: How Long Will the Double Dividend Last?' in T. O'Riordan and J. Jaeger (eds), *The Politics of Climate Change: an European Perspective* (London: Routledge).
Biliouri, D. (1999) 'Environmental NGOs in Brussels: How Powerful are their Lobbying Activities?', *Environmental Politics*, 8, 173–82.
Bobbio, L. and Zeppetella, A. (1999) 'Shifting tools and shifting meanings in urban traffic policy: the case of Turin', in W. Grant, A. Perl and P. Knoepfel (eds), *The Politics of Improving Urban Air Quality* (Cheltenham: Edward Elgar), 73–92.
Boehmer Christiansen, S. (1995) 'Britain and the International Panel on Climate Change', *Environmental Politics*, 4, 1, 1–18.
Bomberg, E. (1996) 'Greens in the European Parliament', *Environmental Politics*, 5, 324–31.
Bomberg, E. (1998) *Green Parties and Politics in the European Union* (London: Routledge).
Bramble, B. and Porter, G. (1992) 'Non-governmental organisations and the making of US international environmental policy' in A. Hurrell and B. Kingsbury (eds), *The International Politics of the Environment* (Oxford: Clarendon Press).
Breckinridge, R.E. (1997) 'Reassessing regimes: The international regime aspects of the Euopean Union', *Journal of Common Market Studies*, 35, 2, 173–89.
Bretherton, C. and Vogler, J. (1997) 'The European Union as an actor in international environmental politics', International Studies Association annual conference, Toronto, 18–22 March.
Brinkhorst, L.J. (1994) 'The European Community at UNCED: Lessons to be drawn for the future' in D. Curtis and T. Henkels (eds), *Institutional Dynamics of European Integration* (Dordnecht: Martinus Nijhoff Publishers).
Brown, Lester (1996) *State of the World: a Worldwatch Institute report* (London: W.W Norton and Co).
Brusco, S., Bertossi, P. and Cottica, A. (1996) 'Playing on two chessboards – the European waste management industry: strategic behaviour in the market and the policy debate', in F. Lévêque, *Environmental Policy in Europe* (Cheltenham: Edward Elgar), 113–42.
Bulmer, S. (1983) 'Domestic politics and European Community policy-making', *Journal of Common Market Studies*, 21, 4, 349–409.
Carson, R. (1962) *Silent Spring* (Harmondsworth: Penguin Books).
Cavender, J. and Jaeger, J. (1993) 'The History of Germany's Response to Climate Change', *International Environmental Affairs*, 5, 1, 3–18.

Cawson, A. (1997) 'Big Firms as Political Actors: Corporate Power and the Governance of the European Consumer Electronics Industry', in H. Wallace and A.R. Young (eds), *Participation and Policy-Making in the European Union* (Oxford: Clarendon Press), 185–205.

CEFIC (1998) *European Council of Chemical Industry Federations: Annual Report* (Brussels: CEFIC).

Cerveny, R. and Balling, R. (1998) 'Weekly cycles of air pollutants, precipitation and tropical cyclones in the coastal NW Atlantic region', *Nature*, 394, 561.

Chatterjee, P. and Finger, M. (1994) *The Earth Brokers: Power, Politics and World Development* (London: Routledge).

Cini, M. (1997) 'Administrative Culture in the European Commission: The Cases of Competition and Environment', in N. Nugent (ed.), *At The Heart of the Union* (Basingstoke: Macmillan), 71–88.

Clifford, M., Clarke, R. and Rifford, S. (1997) 'Drivers' exposure to carbon monoxide in Nottingham', *Atmospheric Environment*, 31, 7, 1003.

Climate Action Network (1995) Independent NGO Evaluations of National Plans for Climate Change Mitigation: OECD Countries and Central and Eastern Europe, 3rd and 1st reviews, Brussels: CNE.

Climate Network Europe (1994) 'Joint Implementation from a European NGO Perspective', Brussels: CNE.

Climate Network Europe (1995) 'Comments on the Commission Green Paper for an European Energy Policy', Brussels: CNE.

Climate Network Europe (1996) Letter to members of the European Parliament on the Draft Directive to introduce Rational Planning Techniques in the Electricity and Gas distribution sectors.

Climate Network Europe (1998a) 'Climate Network Europe', Available: http://www.climatenetwork.org/CNE.html.

Climate Network Europe (1998b) Members' mailing.

Climate Network Europe/Climate Action Network United States (1996a) Independent NGO Evaluation of OECD, Country National Reviews, Fourth Interim Review, Brussels: CNE.

Climate Network Europe/Climate Action Network-United States (1996b) Independent NGO Evaluation of National Plans for Climate Change Mitigation, Fourth (interim) review, Brussels: CNE.

Coen, D. (1997) 'The Evolution of the Large Firm as a Political Actor in the European Union', *Journal of European Public Policy*, 4, 91–108.

Coen, D. (1998) 'The European Business Interest and the Nation-State: Large-firm Lobbying in the European Union and the Member States', *Journal of Public Policy*, 18, 75–100.

Collie, L. (1993) 'Business Lobbying in the European Community: The Union of Industrial and Employer's Confederations of Europe' in S. Mazey and J. Richardson (eds), *Lobbying in the EC* (Oxford: Oxford University Press).

Collier, U. (1993) 'Subsidiarity and Climate Change Policy: an Excuse for Inaction in the European Community', paper presented at the

'Perspectives on the Environment 2: Research and Action' Conference, University of Sheffield, September.

Collier, U. (1996a) 'Implementing a Climate Change Strategy in the EU: Obstacles and Opportunities', European University Institute Working Paper RSC No. 96/1.

Collier, U. (1996b) 'The European Union's climate change policy: Limiting emissions or limiting powers?', *Journal of European Public Policy*, 3, 122–38.

Collier, U. (1997a) 'Sustainability, Subsidiarity and Deregulation: New Directions in EU Environmental Policy', *Environmental Politics*, 6, 2, 1–23.

Collier, U. (1997b) 'Windfall emission reductions in the UK' in U. Collier and R. Löfstedt (eds), *Cases in Climate Change Policy: Political Reality in the European Union*, 87–104.

Collier, U. (ed.) (1998) *Deregulation in the European Union: Environmental Perspectives* (London: Routledge).

Collier, U. (1998a) 'The Environmental Dimensions of Deregulation: an Introduction', in U. Collier (ed.), *Deregulation in the European Union: Environmental Perspectives* (London: Routledge).

Collier, U. (1998b) 'Liberalisation in the Energy Sector: Environmental Threat or Opportunity?', in idem (ed.), *Deregulation in the European Union* (London: Routledge), 3–22.

Collier, U. and Löfstedt, R. (1997) (eds) *Cases in Climate Change Policy: Political Reality in the European Union* (London: Earthscan).

Collins, C. (1996) 'Free Euro Power Market Could Increase Pollution', *Electrical Review*, 30 April–13 May.

Collins, K. and Earnshaw, D. (1992) 'The Implementation and Enforcement of European Community Environmental Legislation', *Environmental Politics*, 1, 213–49.

Committee of Independent Experts (1999) *First Report on Allegations Regarding Fraud, Mismanagement and Nepotism in the European Commission* (Brussels: Commission of the European Communities).

Committee on the Medical Effects of Air Pollutants (1994) *Report May 1992–December 1993 and Advisory Group on the Medical Effects of Air Pollution Episodes Activities Report 1990–1993* (London: HMSO).

Communication from the European Commission (Advance Copy) 'Energy efficiency in the European Community – Towards a strategy for the rational use of energy'.

Conca, K. (1993) 'The environment and the deep structure of world politics' in R. Lipschultz and K. Conca (eds), *The State and Social Power in Global Environmental Politics* (New York: Columbia University Press), 306–27.

Connolly, B., Gutner, T. and Bedarff, H. (1996) 'Organisational inertia and environmental assistance to Eastern Europe' in R. Keohane and M. Levy (eds), *Institutions for Environmental Aid* (Cambridge MA: MIT Press).

Connolly, B. and List, M. (1996) 'Nuclear safety in Eastern Europe and the former Soviet Union' in R. Keohane and M. Levy (eds), *Institutions for Environmental Aid* (Cambridge MA: MIT Press).

Council Decision 94/69/EEC Concerning the Conclusion of the UNFCCC 15 December 1993.
Court of Justice (1988) *Commission* v. *Denmark* [1988], Case 302/86, ECR 4607.
Court of Justice (1992) *Commission* v. *Belgium* [1992], Case C-2/90, ECR I-4431.
Cox, W. (1987) *Production, Power and World Order* (New York: Columbia University Press).
Craig, P. and G. de Burca (1998) *EU Law: Text, Cases and Materials*, Second Edition (Oxford: Oxford University Press).
Crockett, T.R. and Schultz, C. B. (1991) 'The Integration of Environmental Policy and the European Community: Recent Problems of Implementation and Enforcement', *Columbia Journal of Transnational Law*, 29, 169–71.
Dehousse, R., Joerges, C., Majone, G., Snyder, F. and Everson, M. (1992) 'Europe after 1992: New Regulatory Strategies', *European University Institute Working Paper in Law*, 92/31, Florence: EUI.
Dent, C.M. (1997) *The European Economy: The Global Context* (London and New York: Routledge).
Departments of Environment, Health and Transport (1994) 'Health Effects of Particles: The Government's Preliminary Response to the Reports of the Committee on the Medical Effects of Air Pollutants and the Expert Panel on Air Quality Standards', London: Department of the Environment.
Desideri, C. and Lewanski, R. (1999) 'Improving air quality in Italian cities: the outcome of an emergency policy style', in W. Grant, A. Perl, and P. Knoepfel (eds), *The Politics of Improving Urban Air Quality* (Cheltenham: Edward Elgar), 52–72.
DGXI (1997) 'Compliance Costing for approximation of EU environmental legislation in the CEEC', Website, April.
DGXI (1998) 'Implementing Agenda 21 in the European Community 1998 (DGXI website 1998 Agenda 21). Available at: http://europa.eu.int/en/comm/dg11/agend21.htm
Dodds, F. (1997) *The Way Forward: Beyond Agenda 21* (London: Earthscan).
Downs, A. (1972) 'Up and down with ecology: The issue attention cycle', *Public Interest,* 28, 38–50.
Dubash, N.K. and Oppenheimer, M. (1992) 'Modifying the Mandate of Existing Institutions: NGOs' in I. Mintzer (ed.) *Confronting Climate Change* (Cambridge: Cambridge University Press).
EC-Energy Inform (1995) 'Rational Planning Techniques Proposal Finally Adopted by Commission', 31.
The Economist (1992) *'Europe's Industries Play Dirty'*, 9, May 91–2.
ECPI (1998) *European Council of Plasticisers and Intermediates: Annual Report* (Brussels: ECPI).
Eden, S. (1996) *Environmental Issues and Business* (Chichester: John Wiley).
Elsom, D. (1996) *Smog Alert: Managing Urban Air Quality* (London: Earthspan).

ENDS (1996a) 'Energy White Paper Fails to Rise to CO_2 Challenge', 252, 23–4.
ENDS (1996b) 'Lightweight EC Proposal on Energy Efficiency', 253, 42–3.
ENDS (1996c) 'Ministers Agree Weak Standards for Fridge Energy Efficiency', 252, 42.
ENDS (1997a) 'Climate change report exposes Gummer's unambitious target', 265.
ENDS (1997b) 'EC deal on greenhouse gases fails to speed global talks', 266.
ENDS (1997c) 'Energy labels for dishwashers', 268.
ENDS (1997d) 'Ministers agree on climate, vehicle and solvent emissions', 269.
ENDS (1997e) 'Oil industry loses ground in Auto/Oil agreement', 269.
ENDS (1997f) 'Call for tougher EC rules on fridge energy efficiency', 269.
ENDS (1997g) 'Agreement on car efficiency heads for the rocks', 273.
ENDS (1997h) 'Commission produces unambitious strategy on CHP', 273.
ENDS (1997i) 'Energy efficiency agreement on TVs, videos and washing machines' 274.
ENDS (1997j) 'Agreement on energy saving for TVs and VCRs', 275.
ENDS (1997k) 'Commission strategy calls for doubling of renewable energy', 275.
ENDS (1997l) 'Ministers hold fire on climate, reach compromise on landfills', 275.
ENDS (1998a) 'Austrian EU presidency priorities revealed', 21.
ENDS (1998b) Ministers Agree Air Pollution Rules, Lengthy Deadline on Water Quality, London: Environmental Data Services Report, 281, 47.
Environment Policy Reporter (1996) 'EU–Parliament Proposes Changes to IRP Directive', 16, 3.
Environment Watch: Western Europe (1995a) 'EU Mandates Energy Efficiency Labels for Washing Machines, Dryers', 7 July, 8–9.
Environment Watch Western Europe (1995b) 'EU CO_2/Energy Tax Proposal Gets Mixed Reception', 2 June, 13.
Environment Watch Western Europe (1995c) 'Parliament Proposes Tighter Environmental Safeguards in EU Transport Plan, But Also More Roads', 2 June, 13.
Environment Watch Western Europe (1995d) 'Commission Prepares to Relaunch EU Carbon/Energy Tax', 5 May, 9.
Environment Watch Western Europe (1995e) 'Commission Maintains Goal of EU-Wide CO_2/Energy Tax in Medium Term', 19 May 7–8.
Environment Watch Western Europe (1995f) 'EU Presents Proposal for Climate Protocol', 17 November, 10.
Environment Watch Western Europe (1996) 'Doubts over Bjerregaard Dampen EU Policy Expectations for 1996', 5, 1, 1–2.
Environment Watch Western Europe (1997) 'EU energy tax proposal hits national opposition', 21 March, 9.
EPSECC (1998) 'Human Dimensions of Environmental Change' http://www.csv.warwick.ac.uk/PAIS/epsecc.htm

Erlandson, D. (1993) 'The BTU tax experience: What happened and why it happened?' Paper for the Pollution Tax Forum, 19 November.

Eurelectric (1996) 'EP Hearing on Energy Efficiency: Eurelectric's Position on the Proposed Directive on Integrated Resource Planning (IRP)' Press Release, 23 April.

Eurobarometer (1995) *Europeans and the Environment* (Brussels: European Commission).

Europa (1998a) 'Extenal relations'. Available at: http://euopa.eu.int/pol/ext/en/info.htm

Europa (1998b) 'Partnership for integration: Commission presents strategy on integration of environment into other policy areas', 27 May, Brussels. Available at: http://europa.eu.int.

European Citizen The, (1991) No. 8, September.

European Commission (1992a) 'A Community Strategy to Limit Carbon Dioxide Emissions and to Improve Energy Efficiency', COM (92) 246 final, Brussels.

European Commission (1992b) 'Proposal for a Council Directive introducing a tax on carbon dioxide emissions and energy', COM (92) 226 final, Brussels.

European Commission (1993) 'Council Directive 93/76/EEC to limit carbon dioxide emissions by improving energy efficiency', OCJ 237, 28–30.

European Commission (1994) 'Proposal for a Council directive on ambient air quality assessment and management', COM (94) 109 final.

European Commission (1995a) 'A Community Strategy to Reduce CO_2 Emissions from Passenger Cars and Improve Fuel Economy', COM (95) 689 final, Brussels.

European Commission (1995b) 'Climate Change Strategy: A Set of Options' SEC (95) 288 final.

European Commission (1995c) 'Commission Proposal for a Council Directive to Introduce Rational Planning Techniques in the Electricity and Gas Distribution Sectors', COM (95) 369/4, Brussels.

European Commission (1995d) The EC Communication Under the UN Framework Convention on Climate Change, SEC (95), 451 final.

European Commission (1995e) 'Towards Sustainability', Progress Report from the Commission on the Implementation of the European Community Programme of Policy and Action in Relation to the Environment and Sustainable Development, COM (95) 624 final.

European Commission (1995f) *White Paper: An Energy Policy for the EU*, COM (95) 682.

European Commission (1996a) 'A review of the EC strategy paper for reducing methane emissions', COM (96) 557, Brussels.

European Commission (1996b) Communication from the Commission to the Council and the European Parliament: European Community Water Policy, COM (96) 59 final.

European Commission (1996c) Communication from the Commission: Implementing Community Environmental Law, COM (96) 500 final.

European Commission (1996d) 'EP Resolution on the Communication

from the Commission on energy for the future: Renewable sources of energy-Green paper for a Community Strategy', COM (96) 0576, Brussels.
European Commission (1996e) Second Evaluation of National Programmes Under the Monitoring Mechanism of Community CO_2 and Other Greenhouse Gas Emissions Progress Towards the Community CO_2 Stabilisation Target, COM (96) 91 final, Brussels.
European Commission (1997a) Proposal for a Council Directive Establishing a Framework for Community Action in the Field of Water Policy, COM (97) 49 final.
European Commission (1997b) 'Climate change – The EU approach for Kyoto', COM (97) 481, Brussels.
European Commission (1997c) 'Community strategy and action plan for renewable sources of energy', COM (97) 599, Brussels.
European Commission (1997d) Amended Proposal for a Council Directive Establishing a Framework for Community Action in the Field of Water Policy, COM (97) 614 final.
European Commission (1997e) 'Proposal for a Council Directive to introduce rational planning techniques in the electricity and gas distribution sectors', COM (97) 69, Brussels.
European Commission (1998a) Amended Proposal for a Council Directive Establishing a Framework for Community Action in the Field of Water Policy, COM (98) 76 final.
European Commission (1998b) 'An analysis of the Kyoto Protocol', SEC (1998) 467, Brussels.
European Commission (1998c) 'Commission outlines measures to reduce carbondioxide emissions from transport', Brussels.
European Commission (1998d) 'Communication on Transport and CO_2', COM (98) 204, Brussels.
European Commission (1998e) Fifteenth Annual Report on the Implementation of Community Law (1997), COM (98).
European Commission (1998f) 'Partnership for Integration – a Strategy for integrating Environment into European Union Policies', Commission Communication to the European Council, Cardiff.
European Commission (1998g) 'Communication to the Council and the European Parliament, Climate Change: Towards an EU post-Kyoto Strategy', COM (98) 353.
European Communities (1992a) Council Resolution on the Future of Community Groundwater Policy, *Official Journal of the European Communities*, C Series 59, 2.
European Communities (1992b) Declaration on the Implementation of Community Law, *Official Journal of the European Communities*, C Series 191.
European Communities (1993) An European Community Programme of Policy and Action in Relation to the Environment and Sustainable Development: 'Towards Sustainability', *Official Journal of the European Communities*, C Series 138, 17.
European Communities (1996) Commission Proposal for a Ground-

water Action Programme, *Official Journal of the European Communities*, C Series 355, 1.
European Court of Justice (1971) *Commission* v. *Council* (AETR) Case 22/70, ECR 263.
European Environment Agency (1992) *Dobris Report on Europe's Environment*, (Copenhagen: European Environment Agency).
European Environment Agency (1994) *European Rivers and Lakes: Assessment of their Environmental State* (Copenhagen: European Environment Agency).
European Environment (1995) 'Energy Efficiency Requirements for Electric Fridges', 452.
European Environment Agency (1997) *Air Pollution in Europe 1997* (Copenhagen: European Environment Agency).
European Environmental Law Review (1995) 'The Climate Change Convention', June, 190.
European Parliament (1996) Report from the Committee on the Environment, Public Health and Consumer Protection on the Commission Communication to the Council and the European Parliament on European Community Water Policy, *Official Journal of the European Communities*, C Series 347, 52.
European Report (1995) 'Commission Approves Proposal on Rational Resource Planning', 2069, 1–2.
Fee, D. (1995) 'The SAVE II and ALTENER Programmes', *Energy in Europe*, 25.
Flynn, B. (1998) 'EU Environmental Policy at a Crossroads? Reconsidering Some Paradoxes in the Evolution of Policy Content', *European Journal of Public Policy*, 5, 691–6.
Freestone, D.A.C. and Davidson, J.S. (1988) *The Institutional Framework of the European Communities* (London: Croom Helm).
Giddens, A. (1998) 'After the left's paralysis', *New Statesman*, 1 May, 18–21.
Giraud, P.-N., Collier, U. and Löfstedt, R. (1997) 'France: Relying on past reductions and nuclear power' in U. Collier and R. Löfstedt (eds), *Cases in Climate Change Policy: Political Reality in the European Union* (London: Earthscan).
GLOBE (1997) 'Action agenda on global climate change', GLOBE International XIIth General Assembly, 5–7 May, Brussels.
Golub, J. (1996) 'State Power and Institutional Influence in European Integration: Lessons from the Packaging Waste Directive', *Journal of Common Market Studies*, 34, 313–40.
Golub, J. (1997) *The Path to EU Environmental Policy: Domestic Politics, Supranational Institutions, Global Competition*, paper presented at the Fifth Biennial International Conference of the European Community Studies Association, Seattle.
Golub, J. (1998a) 'Global Competition and EU environmental policy: Introduction and Overview' in J. Golub (ed.) *Global Competition and EU Environmental Policy* (London and New York: Routledge).
Golub, J. (1998b) 'New instruments for environmental policy in the EU:

introduction and overview' in J. Golub (ed.) *New Instruments for Environmental Policy in the EU* (London and New York: Routledge).

Grande, E. (1996) 'The state and interest groups in a framework of multi-level decision-making: The case of the European Union', *Journal of European Public Policy*, 3, 3, 318–38.

Grant, W. (1981) 'The Development of the Government Relations Function in UK Firms, a Pilot Study of UK Based Companies', Berlin, International Institute of Management Labour Market Policy Discussion Paper 81/20.

Grant, W. (1993a) 'Pressure groups and the European Community: An Overview' in S. Mazey and J. Richardson (eds), *Lobbying in the EC* (Oxford: Oxford University Press).

Grant, W. (1993b) 'Transnational Companies and Environmental Policy-Making: The Trend of Globalisation' in J. Liefferlink et al (eds), *European Integration and Environmental Policy* (London: Belhaven Press).

Grant, W. (1995) *Autos, Smog and Pollution Control* (Aldershot: Edward Elgar).

Grant, W. (1996) 'Improving Air Quality: Lessons from California' in D. Banister (ed.), *Transport Policy and the Environment* (London: E. & F. Spon).

Grant, W. (1997) *The Common Agricultural Policy* (London: Macmillan).

Grant, W. (1998) 'Large Firms, SMEs, Environmental Deregulation and Competitiveness', in U. Collier (ed.), *Deregulation in the European Union: Environmental Perspectives* (London: Routledge).

Grant, W., Paterson, W.E. and Whitston, C. (1988) *Government and the Chemical Industry* (Oxford: Clarendon Press).

Grant, W., Perl, A. and Knoepfel, P. (eds) (1999) *The Politics of Improving Urban Air Quality* (Cheltenham: Edward Elgar).

Green Cowles, M. (1997) 'Organizing Industrial Coalitions: a Challenge for the Future?', in H. Wallace and A.R. Young (eds), *Participation and Policy-Making in the European Union* (Oxford: Clarendon Press), 116–40.

Green Cowles, M. (1998) 'The Changing Architecture of Big Business', in J. Greenwood and M. Aspinwall (eds), *Collective Action in the European Union* (London: Routledge), 108–25.

Greenpeace (UK) (1994) 'Potential Impacts of Climate Change on Health in the UK', June, London.

Greenpeace (1997) 'Greenpeace exposes European energy subsidy scandal', 20 May.

Greenpeace Business (1993) 'Clean energy industry calls for CO_2 tax', June, London.

Greenpeace Business (1994) 'CO_2 court case: EC funded power stations in jeopardy', February, London.

Greenwood, J. (1997) *Representing Interests in the European Union* (London: Macmillan).

Greenwood, J., Grote, J. and Ronit, K. (1992) 'Conclusions: Evolving patterns of organising interests in the European Community' in J. Greenwood et al. (eds), *Organised Interests and the EC* (London: Sage).

Greenwood, J. et al. (eds) (1992) *Organised Interests and the EC* (London: Sage).
Greenwood, J. and Ronit, K. (1994) 'Interest groups in the European Community: Newly emerging dynamics and forms', *West European Politics*, 17, 1, 31–52.
Grubb, M. (1993) 'EC Climate Policy and the UN Framework Convention on Climate Change' in P. Vellinga and M. Grubb (eds), *Climate Change Policy in the European Community* (London: RIIA).
Grubb, M. (1995) *Global Environmental Problems and the Challenge of Climate Change* (London: RIIA).
Grubb, M. and Brackley, P. (1991) 'Greenhouse responses in the UK and the European Community: Will Britannia waive the rules?' in M. Grubb (ed.) *Energy Policies and the Greenhouse Effect* (London: RIIA/Dartmouth).
Grubb, M. and Anderson, D. (eds) (1995) *The Emerging International Regime for Climate Change* (London: RIIA).
Haas, P. (1990) *Saving the Mediterranean: the Politics of International Environmental Cooperation* (New York: Columbia University Press).
Haas, P. (1998) 'Compliance with EU directives: Insights from international relations and comparative politics', *Journal of European Public Policy*, 5, 4, 17–37.
Haas, P., Keohane, R. and Levy, M. (1993) *Institutions for the Earth: Sources of Effective Environmental Protection* (Cambridge: MIT Press).
Haigh, N. (1989) *EEC Environmental Policy and Britain* (London: Longman).
Haigh, N. (1992) 'The European Community and international environmental policy' in A. Hurrell and B. Kingsbury (eds), *The International Politics of the Environment* (Oxford: Clarendon Press).
Haigh, N. (1995) *Manual of Environmental Policy: the EC and Britain* (London: Longman).
Haigh, N. (1996) 'Climate Change Policies and Politics in the European Community' in T. O' Riordan and J. Jaeger (eds), *The Politics of Climate Change: an European Perspective* (London: Routledge).
Hardi, P. (1994) 'East Central European policy-making: The case of the environment' in O. Höll (ed.), *Environmental Cooperation in Europe* (Boulder CO: Westview Press).
Hawkins, K. (1984) *Environment and Enforcement: Regulation and Social Definition of Pollution* (Oxford: Clarendon Press).
Hayes-Renshaw, F. and Wallace, H. (1997) *The Council of Ministers* (London: Macmillan).
Heller, T. (1998) 'The path to EU climate change policy' in J. Golub (ed.), *Global Competition and EU Environmental Policy* (London: Routledge).
Hession, M. and Macrory, R. (1994) 'Maastricht and the environmental policy of the Community: Legal issues of a new environment policy' in D. O'Keefe and P. Twomey (eds), *Legal Issues of the Maastricht Treaty* (Chichester: Chancery Law Publishing).
Higham, N. (1990) *Marketing Week*, 13 (15), 17.
Hix, S. (1996) 'CP, IR and the EU! A Rejoinder to Hurrell and Menon', *West European Politics*, 19, 4, 802–4.

Houghton, J.T., Jenkins, G.J. and Ephraums, J.J. (eds) (1990) *Climate Change: The IPCC Scientific Assessment* (Cambridge: Cambridge University Press).
House of Lords (1996) 1st Report from the Select Committee on Science and Technology, *Towards Zero Emissions for Road Transport* (London: The Stationery Office).
Huber, M. (1997) 'Leadership in the European climate policy: Innovative policy-making in policy networks' in D. Liefferlink and M.S. Anderson (eds), *The Innovation of EU Environmental Policy* (Copenhagen: Scandinavian University Press).
Hull, R. (1993) 'Lobbying Brussels: a View from Within', in S. Mazey and J. Richardson (eds), *Lobbying in the European Community* (Oxford: Oxford University Press, 1993), 82–92.
Hull, R. (1994) 'The environmental policy of the European Community' in O. Höll (ed.), *Environmental Cooperation in Europe: The Political Dimension* (Oxford: Westview Press), 145–59.
Humphreys, D. (1993) 'The forest debate of the UNCED process', *Paradigms*, 7, 1, 43–55.
Humphreys, D. (1996a) *Forest Politics: The Evolution of International Cooperation* (London: Earthscan).
Humphreys, D. (1996b) 'Regime theory and NGOs: The case of forest conservation' in D. Potter (ed.), *NGOs and Environmental Policies: Asia and Africa* (London: Frank Cass).
Humphreys, D. (1997) 'Hegemonic ideology and the International Tropical Timber Organisation' in J. Vogler and M. Imber (eds), *The Environment and International Relations* (London: Routledge), 215–34.
Hurrell, A. and Menon, A. (1996) 'Politics like any other? Comparative politics, International Relations and the study of the EU', *West European Politics*, 19, 2, 386–402.
Hutter, B.M. (1989) 'Variations in Regulatory Enforcement Styles', *Law and Policy*, 11, 2.
Hutter, C., Keller, H., Ribbe, L. and Wohlers, R. (1995) *The Ecotwisters: Dossier on the European Environment* (London: Green Print).
Ikwue, A. and Skea, J. (1996) 'The energy sector response to European convention emission regulations' in F. Leveque (ed.), *Environmental Policy in Europe* (Cheltenham: Edward Elgar), 31–51.
Jachtenfuchs, M. (1990) 'The European Community and the protection of the ozone layer', *Journal of Common Market Studies*, 28, 3, 261–303.
Jachtenfuchs, M. (1996) *International Policy-Making as a Learning Process* (Aldershot: Avebury).
Jachtenfuchs, M. and Huber, M. (1993) 'Institutional learning in the European Community: The response to the greenhouse effect' in J. Liefferlink et al (eds) *European Integration and Environmental Policy* (London: Belhaven Press).
Jenkins, R. (1989) *European Diary 1977–81* (London: Collins).
Jordan, A. (1998) 'The Ozone Endgame: the Implementation of the Montreal Protocol in the United Kingdom', *Environmental Politics*, 7, 23–52.
Jordan, A. (1999a) 'The Implementation of EU Environmental Policy: a

policy problem without a political solution?', *Environment and Planning C: Government and Policy*, 17, 69–90.

Jordan, A. (1999b) 'European Community Water Policy Standards: Locked In or Watered Down?', *Journal of Common Market Studies*, 37, 113–37.

Jordan, A., Brouwer, R. and Noble, E. (1999) 'Innovative and responsive? A longitudinal analysis of the speed of EU environmental policy-making, 1967–97', *Journal of European Public Policy*, 6, 376–98.

Keohane, R. (1984) *After Hegemony: Cooperation and Discord in the World Political Economy* (Princeton NJ: Princeton University Press).

Klatte, E.R. (1999) 'The Principle of Integration After 25 Years of Community Environmental Policy', *WGES Newsletter*, No. 19, 1–7.

Knox, R. et al. (1997) 'Major grass pollen allergen Lol p1 binds to diesel exhaust particles: implications for asthma and air pollution', *Clinical and Experimental Allergy*, 27, 3, 246.

Koutstaal, P. and Nentjes, A. (1995) 'Tradeable carbon permits in Europe: Feasibility and comparison with taxes', *Journal of Common Market Studies*, 33, 2, 219–35.

Krämer, L. (1991) 'The Implementation of Community Environmental Directives within Member States: Some Implications of the Direct Effect Doctrine', *Journal of Environmental Law*, 3, 1, 39–56.

Krämer, L. (1997) *Focus on European Environmental Law* (London: Sweet and Maxwell).

Krasner, S.D. (ed.) (1983) *International Regimes* (Ithaca NY: Cornell University Press).

Laffan, G. and Shackleton, M. (1996) 'The Budget' in H. Wallace and W. Wallace (eds), *Policy-Making in the European Union*, 71–96.

Larrue, C. and Vlassopoulou, C. A. (1999) 'Changing definitions and networks in clean air policy in France', in W. Grant, A. Perl, and P. Knoepfel (eds), *The Politics of Improving Urban Air Quality* (Cheltenham: Edward Elgar), 93–106.

Lauber, V. (1994) 'The political infrastructure of environmental politics in Western and Eastern Europe' in O. Höll (ed.), *Environmental Cooperation in Europe* (Boulder CO: Westview Press).

Lenschow, A. (1997) 'Variation in EC environmental policy integration: Agency push within institutional structures', *Journal of European Public Policy*, 4, 1, 109–27.

Liberatore, A. (1994) 'Facing Global Warming: The Interactions between Science and Policy-Making in the European Community' in T. Benton and M. Redclift (eds), *Social Theory and the Global Environment* (London: Routledge).

Liberatore, A. (1997) 'The European Union: Bridging domestic and international environmental policy-making' in M.A. Schreurs and E. Economy (eds), *The Internationalisation of Environmental Protection* (Cambridge: Cambridge University Press).

Liefferink, D. and Mol, A.P.J. (1998) 'Voluntary Agreements as a Form of Deregulation? The Dutch Experience', in U. Collier (ed.), *Deregulation*

in the European Union: Environmental Perspectives (London: Routledge).
Liefferink, D. and Andersen, M. (1998) 'Strategies of the "green" member states in EU environmental policy-making', *Journal of European Public Policy*, 5, 254–70
Litfin, K. (1993) 'Eco-regimes: Playing tug of war with the nation-state' in R. Lipschutz and K. Conca (eds), *The State and Social Power in Global Environmental Politics* (New York: Columbia University Press).
Litfin, K. (1994) *Ozone Discourses* (New York: Columbia University Press).
Lowe, P. and Goyder, J. (1983) *Environmental Groups in Politics* (London: Allen & Unwin).
Lowe, P. and Ward, S. (1998a) 'Britain in Europe: themes and issues in national environmental policy' in idem (eds), *British Environmental Policy and Europe* (London: Routledge), 3–30.
Lowe, P. and Ward, S. (1998b) 'Domestic winners and losers' in idem (eds), *British Environmental Policy and Europe* (London: Routledge), 87–104.
Lowi, T. (1964) 'American business, public policy, case studies and political theory', *World Politics*, 16, 4, 677–715.
McCleary, R.M. (1991) 'The international community's claim to rights in Brazilian Amazona', *Political Studies*, 39, December, 691–707.
McConnell, F. (1997) 'The Convention on Biodiversity' in F. Dodds (ed.), *The Way Forward: Beyond Agenda 21* (London: Earthscan), 47–55.
McCormick, J. (1998) 'Environmental Policy: Deepen or Widen?', in P.-H. Kaurent and M. Maresceau (eds), *The State of the European Union, Vol. 4* (London: Belhaven Press).
McLaughlin, A. and Jordan, G. (1993) 'The rationality of lobbying in Europe: why are Euro-groups so numerous and so weak? Some evidence from the car industry?' in S. Mazey and J. Richardson (eds), *Lobbying in the EC* (Oxford: Oxford University Press).
Macrory, R. and Hession, M. (1996) 'The European Community and Climate Change: The Role of Law and Legal Competence' in T. O'Riordan and J. Jaeger (eds), *The Politics of Climate Change: an European Perspective* (London: Routledge).
Maddison, D. and Pearce, D. (1995) 'The UK and global warming policy' in T. Gray (ed.), *UK Environmental Policy in the 1990s* (Basingstoke: Macmillan).
Majone, G. (1996) *Regulating Europe* (London: Routledge).
Mallestone, C. (1981) 'The external relations of the EEC in the field of environmental protection', *The International and Comparative Law Quaterly*, 30, 1, 104–18.
Marchetti, A. (1996) 'Climate change politics in Italy' in T. O'Riordan and J. Jaeger (eds), *The Politics of Climate Change: An European Perspective* (London: Routledge).
Marek, D. (1999) 'Clean air and transport policy in Switzerland: the case of Berne', in W. Grant, A. Perl and P. Knoepfel (eds), *The Politics of Improving Urban Air Quality* (Cheltenham: Edward Elgar), 127–43.

Markham, A. (1994) 'Some like it hot: Biodiversity and the survival of species' (Geneva: WWFI).

Marks, G. and McAdam, D. (1996) 'Social movements and the changing structure of political opportunity in the European Union' in G. Marks, F.W. Scharpf, P.C. Schmitter and W. Streeck (eds), *Governance in the European Union* (London: Sage).

Matláry, J.M. (1996) 'Energy policy: From a national to a European network?' in H. Wallace and W. Wallace (eds), *Policy-Making in the European Union* (Oxford: Oxford University Press).

Matláry, J.M. (1997) *Energy Policy in the European Union* (Basingstoke: MacMillan).

Matthews, D. (1998) 'The Framework Directive on Community Water Policy: a New Approach for EC Environmental Law' in A. Barav and D. A. Wyatt (eds), *Yearbook of European Law 1997* (Oxford: Clarendon Press, published 1998), 191–206.

Matthews, D. and Pickering, J. (1997) 'Directive 80/778 on Drinking Water Quality: An Analysis of the Development of European Environmental Rules', *International Journal of Biosciences and the Law*, 1, 3, 265.

Mayes, D. and Hart, P. (1994) *The Single Market Programme as a Stimulus to Change* (Cambridge: Cambridge University Press).

Mazey, S. and Richardson, J. (1992) 'Environmental groups and the EC: Challenges and opportunities', *Environmental Politics*, 1, 4, 109–28.

Mazey, S. and Richardson, J. (eds) (1993) *Lobbying in the EC* (Oxford: Oxford University Press).

Newell, P. (1996a) 'Climate Politics in Western Europe: Regional and Global Dimensions' Earth Council Report. Available at: http://www.ecouncil.ac.cr/rio/focus/report/english/climate.htm

Newell, P. (1996b) 'Climate change in the Mediterranean: The NGO dimension' Sustainable Mediterranean (Brussels: EC press).

Newell, P. (1997a) 'The International Politics of Global Warming: A Non-Governmental Account', PhD Thesis, Keele University.

Newell, P. (1997b) 'A Changing landscape of diplomatic conflict: The politics of climate change post-Rio' in F. Dodds (ed.) *The Way Forward: Beyond Agenda 21* (London: Earthscan).

Newell, P. (1998) 'Who "CoPed" out in Kyoto? An asessment of the Third Conference of the Parties to the Framework Convention on Climate Change', *Environmental Politics*, 7, 2, 153–60.

Newell, P. and Grant, W. (2000) 'Environmental NGOs and EU Environmental Law', in H. Somsen (ed.) *Yearbook of European Environmental Law, volume 1* (Oxford: Oxford University Press), 225–52.

Newell, P. and Paterson, M. (1998) 'A climate for business: Global warming, the state and capital', *Review of International Political Economy*, 5, 4, 679–703.

O'Riordan, T. (1992) *The Precautionary Principle in Environmental Management*, University of East Anglia, Norwich, CSERGE GEC Working Paper 92–03.

O'Riordan, T. and Jaeger, J. (eds) (1996a) *The Politics of Climate Change: An European Perspective* (London: Routledge).

O'Riordan, T. and Rowbotham, E. (1996b) 'Struggling for Credibility: The United Kingdom's Response' in T. O'Riordan and J. Jaeger (eds), *The Politics of Climate Change: An European Perspective* (London: Routledge).

Page, E.C. (1997) *People Who Run Europe* (Oxford: Clarendon Press).

Parsons, E.A (1993) 'Protecting the ozone layer' in P.M. Haas, R.O. Keohane and M.A. Levy (eds), *Institutions for the Earth: Sources of Effective International Environmental Protection* (Cambridge MA: MIT Press).

Paterson, M. (1992) 'The Convention on climate change agreed at the Rio Conference', *Environmental Politics*, 1, 4, 267–72.

Paterson, M. (1993) 'The Politics of Climate Change after UNCED', *Environmental Politics*, 2, 4, 174–90.

Paterson, M. (1996) *Global Warming and Global Politics* (London and New York: Routledge).

Paterson, M. and Grubb, M. (1992) 'The International Politics of Global Warming' *International Affairs*, 68, 2, 293–310.

Paterson, W.E. (1991) 'Regulatory Change and Environmental Protection in the British and German Chemical Industries', *European Journal of Political Research*, 19, 307–26.

Peters, G.B. (1994) 'Agenda-setting in the European Community', *Journal of European Public Policy*, 1, 1, 9–26.

Peterson, J. and Bomberg, E. (1999) *Decision-Making in the European Union* (London: Macmillan).

Porta, G. (1998) 'Environmental Policy Instruments in a Deregulatory Climate: the Business Perspective', in U. Collier (ed.), *Deregulation in the European Union* (London: Routledge), 165–80.

Presse (1995) Press release of the meeting of the Council of Energy Ministers, 20 December.

Princen, T. and Finger, M. (1994) *Environmental NGOs in World Politics* (London: Routledge).

Putnam, R. (1988) 'Diplomacy and domestic politics: the logic of two-level games', *International Organisation*, 42, 3, 427–60.

Rahman, A. and Roncerel, A. (1994) 'A view from the ground up' in I. Mintzer and J. Leonard (eds), *Negotiating Climate Change* (Cambridge: Cambridge University Press).

Rehbinder, E. and Stewart, R. (1985) *Environmental Protection Policy* (Berlin and New York: de Gruyter)

Richardson, G.M., Ogus, A.I and Burrows, P. (1983) *Policing Pollution: a Study of Regulation and Enforcement* (Oxford: Clarendon Press).

Richardson, J. (1994) 'EU water policy: Uncertain agendas, shifting networks and complex coalitions' *Environmental Politics*, 3, 4, 139–68.

Ringuis, L. (1997) 'Differentation, leaders and fairness: Negotiating climate commitments in the European Community', *CICERO*, 8, University of Oslo.

Risse-Kappen, T. (ed.) (1995) *Bringing Transnational Relations Back In: Non-*

state Actors, Domestic Structures and International Institutions (Cambridge: Cambridge University Press).

Robins, N. (1998) 'Competitiveness, environmental sustainability and the future of European Union development cooperation' in J. Golub (ed.) *Global Competition and EU Environmental Policy* (London and New York: Routledge), 189–212.

Rowbotham, E. (1996) 'Struggling for Credibility: The United Kingdom's Response' in T. O'Riordan and J. Jager (eds), *The Politics of Climate Change: an European Perspective* (London: Routledge).

Rowlands, I. (1995) *The Politics of Global Atmospheric Change* (Manchester: Manchester University Press).

Rowlands, I. (1997) 'EU policy for ozone layer protection', draft of Rowlands (1998).

Rowlands, I. (1998) 'EU policy for ozone layer protection' in J. Golub (ed.) *Global Competition and EU Environmental Policy* (London and New York: Routledge), 34–60.

Rucht, D. (1993) 'Think globally, act locally'? Needs, forms and problems of cross-national cooperation among environmental groups' in D. Liefferlink et al. (eds) *European Integration and Environmental Policy* (London: Belhaven Press).

Saint-Laurent, C. (1997) 'The forest principles and the inter-governmental panel on forests' in F. Dodds (ed.) *The Way Forward: Beyond Agenda 21* (London: Earthscan), 65–83.

Sbragia, A. (1996a) 'Environmental Policy' in H. Wallace and W. Wallace (eds), *Policy-Making in the European Union* (Oxford: Oxford University Press), 235–55.

Sbragia, A. (1996b) 'Environmental policy: The "push-pull" of policy-making' in H. Wallace and W. Wallace (eds) *Policy-Making in the European Union* (3rd edition) (Oxford: Oxford University Press), 235–55.

Sbragia, A. (1997) 'Shaping the institutional architecture of the EU: The influence of global politics', IPSA paper, 17–21 August.

Scott, J. (1998) *EC Environmental Law* (London: Longman).

Seaver, B.M. (1997) 'Stratospheric ozone protection: IR theory and the Montreal Protocol on substances that deplete the ozone layer', *Environmental Politics*, 6, 3, 31–68.

Shaw, J. (1993) *European Community Law* (Basingstoke: Macmillan).

Siedentopf, H. and Hauschild, C. (1989) 'The Implementation of European Union Legislation by the Member States: a Comparative Analysis' in H. Siedentopf and J. Ziller (eds), *Making European Policies Work* (London: Sage).

Siedentopf, H. and Ziller, J. (1989) *Making European Policies Work* (London: Sage).

Skea, J. and Smith, A. (1998) 'Integrating pollution control' in P. Lowe and S. Ward (eds), *British Environmental Policy and Europe* (London: Routledge).

Skjaerseth, J.B. (1994) 'The climate policy of the EC: Too hot to handle?', *Journal of Common Market Studies,* 32, 1, 25–45.

Skolnikoff, E. (1990) 'The policy gridlock on global warming' *Foreign Policy*, 79, 77–93.

Smith, S. (1993) 'The environment on the periphery of international relations: an explanation', *Environmental Politics*, 2, 4, 28–45.

Snyder, F. (1990) *New Directions in European Community Law* (London: Weidenfeld and Nicolson).

Solomon, J. (1998) 'To Drive or to Vote?', Chartered Institute of Transport discussion paper, London.

Somsen, H. (1996) 'The European Union and the OECD' in J. Werksman, (ed.), *Greening International Institutions* (London: Earthscan Publications).

Stairs, K. and Taylor, P. (1992) 'Non-Governmental Organisations and the Legal Protection of the Oceans: a Case Study' in A. Hurrell and B. Kingsbury (eds), *The International Politics of the Environment* (Oxford: Clarendon Press).

Susskind, L. (1994) *Environmental Diplomacy: Negotiating More Effective Global Agreements* (Oxford: Oxford University Press).

SustainAbility (1997) *Food Integrity* (London: SustainAbility).

Sutherland, P. et al (1992) 'The Internal Market After 1992. Meeting the Challenge.' Report to the EEC Commission by the High Level Group on the Operation of the Internal Market, Brussels.

Temple, L.J. (1986) 'The ozone layer convention: A new solution to the question of community particiaption in "mixed" international agreements', *Common Market Law Review*, 23, 157–76.

Thairs, E. (1998) 'Business lobbying on the environment: the perspective of the water sector', in P. Lowe and S. Ward (eds), *British Environmental Policy and Europe* (London: Routledge), 153–70.

Transport and Environment Europe (1998) 'Commission and car industry CO_2 agreement: A good deal or a move backwards?', Press release, 28 July.

Ungar, S. (1992) 'The Rise and (Relative) Decline of Global Warming as a Social Problem', *The Sociology Quarterly*, 33, 4, 483–501.

United Kingdom Photochemical Oxidants Review Group (1993) *Ozone in the United Kingdom 1993* (London: Air Quality Division, Department of the Environment).

Vellinga, P. and Grubb, M. (1993) *Climate Policy in the European Community* (London: RIIA).

Vickerman, R. (1998) 'Transport Provision and Regional Development in Europe', in D. Banister (ed.) *Transport Policy and the Environment*, (London: E. and F.N. Spon), 131–60.

Villot, X.L (1997) 'Spain: Fast growth in CO_2 emissions' in U. Collier and R. Löfstedt (eds), *Cases in Climate Change Policy: Political Reality in the European Union* (London: Earthscan).

Vogel, D. (1986) *National Styles of Regulation* (Ithaca: Cornell University Press).

Vogel, D. (1996) 'The Making of EC Environmental Policy', in M. Ugur (ed.), *Policy Issues in the European Union* (Dartford, Kent: Greenwich University Press), 121–34.

Vogel, D. (1997) 'Trading up and governing across: Transnational governance and environmental protection', *Journal of European Public Policy*, 4, 4, 556–71.

Vogel, D. (1998) 'EU environmental policy and the GATT/WTO' in J. Golub (ed.) *Global Competition and EU Environmental Policy* (London and New York: Routledge), 142–61.

Vonkeman, G. (1996) 'International cooperation: The European Union' in A. Blowers and P. Glasbergen (eds), *Environmental Policy in an International Context: Prospects for Environmental Change* (London: Arnold), 105–33.

Wagner, J.P (1997) 'The climate change policy of the European Union' in G. Fermann (ed.), *International Politics of Climate Change: Key Issues and Critical Actors* (Oslo: Scandinavian University Press).

Wallace, H. (1996) 'Politics and Policy in the EU: the Challenge of Governance', in H. and W. Wallace (eds), *Policy-Making in the European Union* (Oxford: Oxford University Press).

Wapner, P. (1995) 'Politics beyond the state: Environmental activism and world civic politics', *World Politics*, 47, 311–40.

Warren, A. (1993) 'Energy Efficiency Policy in the European Community: How CO_2 Emissions May (or May Not) be Reduced' in P. Vellinga and M. Grubb (eds), *Climate Change Policy in the European Community* (London: RIIA).

Warren, A. (1996) '"Derisory" Funding for Key Strategy', *Energy in Buildings and Industry*, June, 8.

Weale, A. (1992) *The New Politics of Pollution* (Manchester: Manchester University Press).

Weale, A. (1996) 'Environmental Rules and Rule-Making in the European Union', *Journal of European Public Policy*, 3, 594–611.

Weale, A. and Williams, A. (1992) 'Between Economy and Ecology? The Single Market and the Integration of Environmental Policy', *Environmental Politics*, 1, 4, 45–64.

Weiler, J. (1991) 'The Transformation of Europe', *Yale Law Journal*, 100, 2463–4.

Weir, F. (1996) Former Climate Campaigner, Friends of the Earth UK, Questionnaire on The Politics of Global Warming in P. Newell (1997) *The International Politics of Global Warming: A Non-Governmental Account*, PhD Thesis, Keele University.

Whitelegg, J. (1994) 'Road-Builders Make their Pitch', *New Scientist*, 30 April, 48–9.

Wilkinson, D. (1997) 'Towards sustainability in the European Union? Steps within the European Commission towards integrating the environment into other European Union policy sectors', *Environmental Politics*, 6, 1, 153–74.

Woodruff, T. et al. (1997) 'The relationship between selected causes of postneonatal infant mortality and particulate pollution in the United States', *Environmental Health Perspectives*, 105, 6.

World Wide Fund for Nature (1996) 'Intensifying efforts on the Berlin Mandate', June, position statement.

Wurzel, R.K.W. (1996) 'The Role of the EU Presidency in the Environmental Field: Does it Make a Difference which Member State Runs the Presidency?', *European Journal of Public Policy*, 3, 272–91.

Wynne, B. (1993) 'Implementation of greenhouse gas reductions in the European Community: Institutional and cultural factors' *Global Environmental Change*, 3, 1, 101–28.

Wynne, B. and Waterton, C. (1998) 'Public information on the environment: the role of the European Environment Agency', in P. Lowe and S. Ward (eds), *British Environmental Policy and Europe* (London: Routledge), 119–37.

Zito, A.R. (1995) 'Integrating the environment into the European Union: The history of the controversial carbon tax' in C. Rhodes and S. Mazey (eds), *The State and the European Union: Building a European Polity?* (Essex: Longman).

Index

Also by Wyn Grant

AUTOS, SMOG AND POLLUTION CONTROL

BUSINESS AND POLITICS IN BRITAIN

THE COMMON AGRICULTURAL POLICY

GOVERNMENT AND INDUSTRY: A Comparative Analysis

PRESSURE GROUPS AND BRITISH POLITICS

THE POLITICS OF ECONOMIC POLICY

THE POLITICAL ECONOMY OF CORPORATISM (*editor*)

THE POLITICAL ECONOMY OF INDUSTRIAL POLICY

Also by Duncan Matthews

GLOBAL BUSINESS, TRADE AND INTELLECTUAL PROPERTY LAW: The TRIPS Agreement

Also by Peter Newell

CLIMATE FOR CHANGE: Non-State Actors and the Global Politics of the Greenhouse

The Effectiveness of European Union Environmental Policy